Lilian Bell

The Instinct of Step-Fatherhood

Lilian Bell

The Instinct of Step-Fatherhood

ISBN/EAN: 9783337365424

Printed in Europe, USA, Canada, Australia, Japan

Cover: Foto ©berggeist007 / pixelio.de

More available books at **www.hansebooks.com**

The Instinct of Step-Fatherhood

By LILIAN BELL

Author of "A Little Sister to the Wilderness" etc.

HARPER & BROTHERS PUBLISHERS
NEW YORK AND LONDON
1898

By LILIAN BELL.

THE LOVE AFFAIRS OF AN OLD MAID. 16mo, Cloth, Ornamental, Uncut Edges and Gilt Top, $1 25.

... The love affairs of an old maid are not her own, but other people's, and in this volume we have the love trials and joys of a variety of persons described and analyzed.—*Cincinnati Commercial-Gazette.*

THE UNDER SIDE OF THINGS. A Novel. 16mo, Cloth, Ornamental, Uncut Edges and Gilt Top, $1 25.

... This book is Miss Bell's best effort, and most in the line of what we hope to see her proceed in, dainty and keen and bright, and always full of the fine warmth and tenderness of splendid womanhood.—*Interior*, Chicago.

FROM A GIRL'S POINT OF VIEW. With a Photogravure Portrait. 16mo, Cloth, Ornamental, $1 25.

Will be read with zest by every one who loves a crisp and graceful writing.... Altogether this little sheaf of half-whimsical, half-humorous essays will afford as delightful an hour of reading as any person, not dull, can desire.—*Brooklyn Eagle.*

A LITTLE SISTER TO THE WILDERNESS. A Novel. New Edition. 16mo, Cloth, Ornamental, $1 25.

The writer is well acquainted with the life and habits and dialect of the West Tennessee bottoms, and her story is written from the heart and with rare sympathy. It is valuable because it shows so forcefully the peculiar phases of the life and human character of these people.—*Churchman*, N. Y.

NEW YORK AND LONDON
HARPER & BROTHERS, PUBLISHERS.

Copyright, 1898, by HARPER & BROTHERS.

All rights reserved.

TO

MY LOVELY MOTHER

whose descent, not only from the Pilgrim Fathers, but from Plymouth Rock, enables her to withstand my frantic appeals when she is minded too ruthlessly to clip the wings of the fledgelings of my heart and brain.

Nevertheless, to her extraordinary critical faculty do I owe much of the gentle criticism of the public, and to her unfailing tenderness and patience do I hereby publicly bear witness.

CONTENTS

	PAGE
THE INSTINCT OF STEP-FATHERHOOD	3
A STUDY IN HEARTS	25
THE HEART OF BRIER ROSE	75
LIZZIE LEE'S SEPARATION	99
MARY LOU'S MARRYIN'	125
THE STRIKE AT THE "BILLY BOWLEGS"	169
A WOMAN OF NO NERVES	203

THE
INSTINCT OF STEP-FATHERHOOD

THE INSTINCT OF STEP-FATHERHOOD

They had been driving along the big road for an hour, Miss Cornelia and Fred Yarborough, Miss Cornelia studying her bank-book with a perplexed brow, and Fred watching her with a funny imitation of her anxiety on his small square countenance. And finally, when Miss Cornelia closed her book with a deep sigh, Fred fetched one so much deeper and more lugubrious that she turned and looked at him.

"What is the matter, Fred?"

Fred cast a comprehensive glance at the bank-book held daintily in one of Miss Cornelia's small hands, and said, "Livin' is powerful expensive, ain't it, Miss C'nelia?"

"Well, I shouldn't think *you* would find it so, Fred."

Miss Cornelia smiled slowly. Fred generally smiled when Miss Cornelia did, but this time he not only failed to respond but sighed again. For some moments they drove along in silence, Fred occupied with his favorite pastime of contemplating the stubby sunburned hands which held the reins, and comparing them with Miss Cornelia's. Of course Fred knew—everybody knew—that the family at Arborvitæ were famous for the symmetry and beauty of their hands and feet, while Fred Yarborough was man-of-all-work for Miss Cornelia and Miss Caroline, and came of a family whose physical attributes seemed to have been measured by a carpenter's rule, so square, so flat, so uncompromising they were.

"Well-um, I do. I'm just lavishin' money awn myseff these days, Miss C'nelia. I paid a dollar and a half for these very shoes, and I owe Mist' Tarbell three dollars for a coat that I 'ain't never wo'e. I—I'm savin' it."

INSTINCT OF STEP-FATHERHOOD

Fred colored so through his freckles that Miss Cornelia noticed it.

"Well, it's for you to say whether you can afford these extravagances, Fred."

"Well-um, I'm saving in other *di*-rections. I've quit chawin'. It ain't right for a man to chaw when he knows he is settin' a bad example for childern."

Fred's steady blue eyes met Miss Cornelia's with an expression of manliness in their depths which struck her as new. He was only seventeen.

"But all your mother's children are girls, Fred."

"I know it, Miss C'nelia. I know it mighty well."

"How many are there now, Fred?"

"I got 'leven little sisters, Miss C'nelia. Yes'm. All girls. I'd give a heap if a few of 'em was only boys. I just nachally love boy babies. Looks like the mo' I want boys the mo' girls is sent to us. We got so many, Miss C'nelia, I come mighty near bein' sick of girls. 'Pears like we ought to have had one of 'em boys, just for a relish. But Maw, she don't care. She don't seem to hanker after boys like I do. She's just as well satisfied to see 'em all girls. But then Maw don't care much *what* happens. She, being a woman, don't seem to be able to sense her 'sponsibilities. But a time comes when a man must take awn other obligations besides them that he was born into. I don't never intend to turn my back entirely awn Maw, but I've quit chawin' just the same, Miss C'nelia."

Miss Cornelia looked puzzled.

"I am glad to hear it. But, Fred, I hope you are not courting any of these girls around here. Remember, you are only getting eight dollars a month, and that loft in the barn is no place for a bride."

Fred laughed in a pleased way.

"Laws, no, Miss Nely. I ain't studyin' 'bout no *girl*."

As a Southern woman, Miss Cornelia knew that at seventeen it was none too early to warn a boy from matrimony. Many a one is well settled down and the father of a family at nineteen.

They drove in at the gate of Arborvitæ, the fine old place of Judge Peebles, dead

INSTINCT OF STEP-FATHERHOOD

and gone these many years, which was with difficulty kept up by his two spinster daughters.

Deep fields spread far away to the right and left of the famous arbor-vitæ hedge, whose length and luxuriance gave the place its name. Back of the house, with its great porch and massive doors, which in years gone by swung open with a hospitality second to none, the peach orchard stretched out richly. The barn and stable too, although guiltless of the blooded stock they once boasted, were well kept, and helped to maintain the dignity which the fine old place always impressed upon the beholder.

But, as if there must be some blot upon such a picture, some one slattern element to mark both the decay and the carelessness of the South, the skeleton of an old buggy stood, year in and year out, in the stable-yard, in full view both from the drive and from the dining-room windows.

Three times a day, at least, must Miss Cornelia and Miss Caroline view its tattered and flapping remains. The shafts were broken, the hinges rusted, the curtains torn,

INSTINCT OF STEP-FATHERHOOD

and the dash-board was only a weeping apology for itself. Every time the wind rose in the night the small window-panes in the tattered curtains dashed their metal rims against the skeleton frame, and the rusted springs squeaked and shook. Yet no one ever spoke of it disparagingly. No one ever had suggested removing it. And Miss Cornelia, gazing about her with pride, did not notice it now.

As she went in she heard voices, and, walking through the great hall which divided the house into two equal parts, she found Fred's mother on the back porch in earnest conversation with Miss Caroline.

The Yarboroughs were such an improvident set that without the continued help of the ladies of Arborvitæ they must have died from sheer inability to feed themselves.

"Good-evenin', Miss C'nelia. I just come up to Arborvity to ask you 'n' Miss Calline if you'd seen any signs of Fred's marryin'. I been suspicionin' for 'bout three months now that Fred was studyin' 'bout marryin', an' when he never brought his clo'es home to be washed last week I reck-

INSTINCT OF STEP-FATHERHOOD

oned he was fixin' to run off. You say you 'ain't seen no signs of such *in*-tentions, Miss Calline?"

Miss Caroline disappeared for a moment, and returned carrying a small black oil-cloth satchel about eighteen inches long.

"Not unless you call this a sign," she said.

"Yessum, that's it. I knew it. I've seen her with that very baig. She always carries it to camp-meetin' with her to hold the children's clo'es."

"The children's clothes!" cried Miss Cornelia and Miss Caroline at once.

"Yessum. It's the Widow Perkins he's a-courtin'. She's got fo' boys. He's undertakin' right smart of a family to begin with, an' I'm sot agin it. 'Tain't that I would begrudge a pore lone widow woman a husband, but I do hate to see her catch Fred. She's sent after his clo'es an' washed 'em herseff, an' they're fixin' to marry. You wait an' see if they ain't. Any other signs, Miss C'nelia?"

"He asked me if he might have the old buggy to-morrow afternoon," said Miss Cornelia.

"I thought so! You see, Miss Calline! Any other signs?"

"He showed me his new shoes just now," said Miss Cornelia.

"Shoes!" cried Mrs. Yarborough. "He *never* had awn them expensive shoes! An' to-day only Wednesday! Oh, his extravagance will be the death of me!"

She rose to go.

"She's tried so mortal hard to catch him, an' put out so many arts, I do hate to see her *de*-vices succeed," she said. "Fred's j'ined the church awn probation, but I don't believe that's her doin', for she ain't a perfessor. I don't know what he j'ined for."

"Will you take this satchel?" asked Miss Caroline.

"No, 'm; no, 'm. I wouldn't tetch it. That's one of her *de*-vices to catch Fred with. I wouldn't tetch it."

Mrs. Yarborough went slowly down the steps. It did not seem to occur to her to try to prevent this ill-starred marriage, nor did she seem alarmed to know that her chief income of eight dollars a month would go from her to Fred's bride. She felt that the

INSTINCT OF STEP-FATHERHOOD

Lord and Miss Caroline would provide. They always had.

They watched her stubby figure waddle away.

"To think of it!" said Miss Cornelia. "What a burden the child is taking on himself! Bertha Perkins is thirty if she is a day, poor as she can be, and has those four boys, the youngest a mere baby. Her husband has only been dead about eight months."

"You don't think for one moment, Sist' Nely, that I am going to allow it, do you?"

Miss Caroline was the firm one. Miss Cornelia was the one people loved.

"I hadn't thought at all, honey, or I should have known that you meant to stop it."

"That woman is just too lazy for any use. She never would work, even when she was a girl. She just sits and rocks all day long, and she is getting so big and fat she is a sight."

That night, after Miss Caroline had gone to bed, Aunt Easter, the black cook, put an

INSTINCT OF STEP-FATHERHOOD

ashen face in at the door, and, with chattering teeth, said:

"Del Law, Miss Nely, dis place done ha'nted, sho'! Jis creep out in de kitchen one minute an' hyah dat chile cryin'. Hyah dat? An' you know dey ain't a baby within fo' miles of Arbohvity. Dat means a death, sho'. Del Law, Miss Nely, you reckon hit mean me? I ain't fitten to die, Miss Nely. I ain't never perfessed 'ligion. Del Law! Hyah dat? But if de good Lawd spah my life tell nex' Sunday I gwine be babtized in de Branch, sho's shootin'. I is, foh a fack. Oh, Lawd, spah me, hab mussy awn my sinful soul! Hab mussy, Lawd, an' plenteous redemtion, lake You done promus in de Good Book. Del Law, Miss Nely, you hyah dat? Miss Nely, you's a good lady, *an'* a perfessor; won't you jis kneel down awn de flo' an' say one prah foh po' ole Easter? Oh, Lawd, Miss Nely!"

"Hush, Easter," said Miss Cornelia; "that is a real child crying. It is no ha'nt. Give me the lantern and come with me."

Miss Cornelia lifted her skirts and stepped down the path leading to the barn,

INSTINCT OF STEP-FATHERHOOD

in whose upper window she saw Fred's light burning. Aunt Easter, with pious ejaculations of "Don't strike, Lawd!" "Hol' Dy wrath a little longer, Lawd!" followed, holding her clothes up with both hands, and displaying feet and ankles never intended for rapid transit. Nevertheless, Aunt Easter meant to be ready to run if "the ha'nt" appeared, for at each cry of the child she took a fresh grab at her skirts and raised them higher, with frank confidence in the friendly darkness.

When they reached the barn, Miss Cornelia's fears were realized. They could hear Fred's sturdy bare feet patting across the floor and the crying of a child being jolted. Miss Cornelia was in a quandary. She had no doubt but that Fred had married Mrs. Perkins that day and had brought his interesting family home to share the luxury of his loft with him. She did not know what to do. Miss Caroline would have marched up and demanded to know the truth, and Miss Cornelia knew that in the morning she would be held accountable for not taking the same summary steps. She

INSTINCT OF STEP-FATHERHOOD

thought to temporize by going to the door and calling to Fred through the key-hole. But at the first sound of her slippered feet on the stairs, Fred's door flew open and he appeared in his shirt-sleeves, holding Mrs. Perkins's baby in his short arms.

"What's the matter, Miss Nely?"

"What in the world, Fred!" they said at the same time.

"It's the baby, Miss Nely. His Maw was clean wo'e out with tendin' him, an' he takes to me so mightily, I jes' 'lowed I'd take care of him one night an' give her a chance to sleep."

"Where is she?" asked Miss Cornelia.

"She's home!"

"Then you aren't married, Fred?"

"No, 'm, not yet."

It was so ridiculous that Miss Cornelia wanted to laugh, but Fred's face was so grave and the baby so pretty that something of the other side of the situation struck across Miss Cornelia's consciousness. Fred seemed to have swept and garnished his room amazingly. Miss Cornelia sat down in one chair and Fred sat in the oth-

INSTINCT OF STEP-FATHERHOOD

er — there were but two — while he explained.

"You see, Miss Nely, I'm used to babies. Maw's had 'leven besides me — all girls — and I've hepped her with every one of 'em, an' I know jes' what to do. Now Mis' Perkins, altho' she's had fo', an' she bein' their own borned mother, she don' rightly know how to keep a chile from hollerin' and yellin'. So it come about first with them takin' to me so powerful, an' me bein' so fond of the little fellers — all boys, Miss Nely. Then I reckon she saw how handy I'd be to have around, for she kind of took to me herseff. You know she is big and dark, an' I'm little an' fair, an' it do look like opposites like us was just nachally drawed together an' 'pinted to marry. She's lonesome, Miss Nely, an' she ain't wealthy, an' she ain't strong, so she needs somebody to take care of her. No, 'm, she don't look delicate, but she is. She hates to set in a straight cheer. She likes a rocker. I got that one you're settin' in for her. It cost me a dollar and seventy-five cents, but I don't begrudge it to her, Miss Nely."

INSTINCT OF STEP-FATHERHOOD

The baby cried just here, and Fred got up and began to walk up and down with it with a dignity which sat funnily upon his short body.

"Is—is he teething?" asked Miss Cornelia, awkwardly.

"No, 'm. Leastways that ain't what makes him cry. She's weanin' him, an' I 'lowed it would be better to have him tooken clean away from her, where he couldn't see nor hear her. You know she's only got two rooms in her house. So that's what I'm doin'. I'm weaning him for her. I 'ain't never weaned only girl babies befo', Miss Nely, but this un's a boy. An' bein' a boy, an' liable to grow up to bad ways is where the 'sponsibility comes in, Miss Nely, of a man takin' awn family cares. I wouldn't love to have these little fellers know their Paw had no bad habits, Miss Nely. Marryin' is expensive, I know, but Maw will jes' have to give up some of her luxuries, an' git along awn a dollar a month from me, or maybe two. I mus' have at least six dollars a month for my own family. There, Miss Nely, jes' come an'

INSTINCT OF STEP-FATHERHOOD

look at him. Didn't he go to sleep pretty?"

Miss Cornelia and Aunt Easter crept back to the house as softly as they had come. All Miss Cornelia's fine intentions of talking Fred out of his intended marriage had faded into thin air before his earnest little square face and the manliness of his honest eyes. The grotesqueness of the situation did not seem as absurdly patent as before.

"After all, what matter many years or few, when truth and loyalty and chivalry meet together in one sturdy little heart?" thought Miss Cornelia, with a sentimentality which Miss Caroline would have pooh-poohed.

The next day was such a busy one, on account of making the preserves—watermelon-rind preserves, which everybody knows is fussy work, and tomato preserves, done with such care that each pear-shaped tomato would hold its shape and yet be dripping in its own luscious syrup (Miss Caroline always used the pound-for-pound recipe for her preserves)—that it was the middle of the

INSTINCT OF STEP-FATHERHOOD

afternoon before they noticed that both Fred and the old buggy were gone.

Miss Caroline was visibly agitated. Her delicate fingers were shrivelled with alum-water and stained with fruit as they trembled before Miss Cornelia's stricken gaze. Miss Caroline knew nothing of "the ha'nt" of the night before.

"He has gone! And he will be married before we can stop it now! Oh, that foolish child! Sist' Nely, why didn't you make me talk to him last night?"

"Yonder dey come!" cried Aunt Easter. "I sees de ole buggy awn de brow ob de hill. De weddin' pahty's awn de way!"

"Then they must pass here. Sist' Nely, come with me down to the big road—yes, just as you are. This is a call of duty."

She took her sister's reluctant hand, and both old ladies, with the other hands neatly holding their skirts from feet still trim and small, stepped down the arbor-vitæ drive to the big road, down whose smooth stretch came the Yarborough gray mare hitched to the sad old skeleton buggy of the Peebles.

It was a sorry-looking wedding procession. The wheels of the buggy groaned under the weight of the placid, large, mild-eyed woman who towered above her small but earnest bridegroom. The baby was in her lap, the oldest boy on the seat between them, and the other two packed in at their feet. The torn curtains flapped in the breeze and gave a certain banner effect to an otherwise sombre equipage, and through the holes in the dash-board the lively youngsters pulled the patient old mare's tail and prodded her with little sticks.

Mrs. Perkins was not at all abashed to be stopped in this manner. She listened placidly to all Miss Caroline had to say as to why this marriage should not go forward, but Fred's appealing blue eyes were glued to Miss Cornelia's, for in her he felt that he had an ally. She finally felt obliged, in the face of her sister's convincing arguments, to shake her head at him.

"Did I understand you to say, Miss Calline, that if Fred persists in his *in*-tentions of marryin me that you won't let him work for you no mo' ?"

"You certainly did, Bertha; so the sooner

you turn around and go home the better for you."

"Fred, honey," said the bride-elect, "it ain't no use. If that eight dollars a month goes, how you goin' to pervide for a family? He's so taken with the childern, Miss Calline, that I do hate to disapp'int him; but if yore mind is made up, I know mightily well it ain't no mortal use trying to git you to change it. Fred, honey, you know I tole you I reckoned I'd better marry Mist' Tarbell, but you couldn't seem to let them childern go to no man who would bring 'em up every which a-way. Come awn, honey, Miss Calline's done spoke the final word, so let's give up peaceable."

Fred fetched another sigh, such as always accompanied any thought of his approaching marriage, and slowly turned the mare's head.

Nobody had made any fuss. It was all very calm and proper. The two ladies watched the wobbly course of the disappointed wedding-party as it crept back up the hill and disappeared over its brow.

That night Miss Cornelia saw Fred sit-

INSTINCT OF STEP-FATHERHOOD

ting disconsolately alone, and went out to him.

"I am so sorry, Fred, for your sake," she began.

"Nemmine me, nemmine me, Miss Nely," he said. "But think of them little fellers with that man Tarbell for a Paw. He ain't a perfessor nor nothin'."

He sighed deeply.

"They sho' were sweet little fellers, weren't they, Miss Nely? An' all boys. I tell you, Miss Nely, it would a' been worth while raisin' a family like that to be good an' honest men. It used to make me feel powerful solemn just thinkin' about it."

"To be good an' honest men," he repeated.

He smiled up into Miss Cornelia's face.

"It was them I give up chawin' fur, Miss Nely," he said, softly.

A STUDY IN HEARTS

A STUDY IN HEARTS

I.

When an attractive American girl is bored, it generally means that she is not in love with any one. It never means that no one is in love with her. That unfortunate state of things would cause her to be discontented — not bored. Besides, there is always *somebody* in love with the attractive American girl. Unhappily it too often is the wrong somebody.

Jessica Drew was bored.

She sat by the window watching for the postman. Not that she was looking for any particular letter, but when several men are interested in a girl, the advent of the postman is always fraught with mildly exciting possibilities.

There is something particularly ruminative about the occupation of watching for the postman. A woman is liable to feel gently sentimental at such a time. There are so few things in this rushing American life of ours which tend that way, that for this reason, perhaps, the habit of watching for the postman should be encouraged.

Jessica was feeling a trifle hurt — just enough so to make her thoughts amiably cynical. She divided girls into three classes. If a girl had one lover, she was called "a sweet creature" by the other girls. If she had two or three, she was respectfully alluded to as "fascinating." If she were unhappy enough to have won half a dozen, with more in prospect, she was stigmatized as "a coquette."

Jessica had just been called a coquette.

Is it so short a step from the sentimental to the moral, or why is it that conscience often utilizes the quiet moments of watching for the postman to do its most effective work?

Jessica and her conscience were on intimate but not particularly agreeable terms.

One sometimes has friends of that description.

In analyzing why she was neither entirely pleased nor entirely offended by the epithet, Jessica discovered, in one of those sudden blinding soul-illuminations, in which one's conscience takes such ghoulish delight, that her vanity was pricked at the power and fascination over men which the term implied, but that her higher nature revolted at the implication that she would stoop to trifle with love.

In her secret heart she knew that sometimes she did so stoop. But never at the beginning of an affair, she assured herself. It never would occur to her to try to win a man's love. But when she discovered that he already was entangled, why, then— why, then—perhaps she led him on—a trifle. It was wrong, of course. So many agreeable diversions *are* wrong. But how in the world, she argued, warmly, are you going to find out whether you like a man unless you do encourage him? You never even begin to know him until he falls in love with you!

Jessica's cheeks flushed. Then her eyes

grew wistful. To be sure, she had expected that when love came to her heart he would herald his approach with no uncertain sound. But as he never had come thus far, she felt justified in experimenting a trifle to test her own feelings, somewhat regardless of her lover's—whoever he happened to be.

She knew so well that she had no right to do this under any pretext that her conscience became even more disagreeable as a companion than usual.

She wished that postman would hurry! There was no sense in his stopping to talk with all the housemaids along the street!

She comforted herself a little with the thought that she alone knew the truth concerning her shortcomings, for she possessed to a degree the tact of compelling from others a faith in her goodness which she in nowise shared with them. The men themselves always went away with the idea that if Fate had not been adverse, Jessica undoubtedly would have married them.

Thus she was spared the reproaches of her lovers. They even defended her against her women friends, the unmarried of whom were not so considerate.

She reformed periodically. But to her dismay she discovered that the only result therefrom was to cause her to be brusque or in some way unpleasant to an incipient lover, who, as yet, not being sufficiently under her spell to forgive her manner as a vagary of perfection, thereupon became offended and withdrew, under the impression that she either was inordinately conceited, or else had a very bad temper. So, smarting under this palpable injustice, she swung back to her habit of letting things take their own course.

She called her young ladyhood " a career of magnificent beginnings and reprehensible endings."

Her affairs were singularly subjective, for in spite of this " weakness "—her own word —her ideals of womanliness were lofty. Men might pour out their hearts to her in language which would have broken down a reserve less fixed than hers, but they were not permitted so much as to touch the tips of her fingers. Her personal dignity kept at bay all profanation save that of words. In this way she compelled their respect, and kept it after they were released from the spell of her magnetic personality.

An affair of the heart of this type can only be brought to perfection by the American girl, and it has no name. Jessica objected to the word flirtation as only a degree less vulgar than the deed. She considered "coquette" frivolous, and claimed that the French word carried with it no idea of the brain necessarily involved in a woman who essayed the rôle in America.

The charm of Jessica Drew defied analysis. She fully appreciated the fascination with which a certain amount of mystery surrounds a woman. Her manner carried out a mysteriousness which her face, with its strange lights and shadows, suggested. One felt at once that she was singular; then attractive; then alluring. But one must withdraw from her presence to discover that under it all she was supremely clever; so clever even as to conceal her cleverness from men. She labelled herself "Harmless." Men oftener found her deadly.

A STUDY IN HEARTS

As is the case with a masterful personality, an unusually shrewd intellect, and an inflexible will (though in the case of Jessica all this was sheathed under a guileless exterior), she had hitherto attracted men of the opposite temperament. The gentle, considerate, even-tempered man, with no idea of controlling circumstances or braving Fate, always found himself sooner or later at the feet of Jessica Drew. One cannot be sure that she ever would have called the apathy of feeling which this type of man created in her by so harsh a name as contempt, but certain it is that those mysterious eyes of hers had the suspicion of a look in them which was almost mocking, and her restlessness sometimes made itself known in an occasional satirical speech, directed with fatal precision at the weakest point of her lover's armor—a speech barbed with a wit which almost brought the water to his eyes, but which, uttered in her velvet voice, he never had the courage nor the skill to parry.

"A letter for you, Jessica," said a voice at the door.

She turned with a start. That insidious

postman must have evaded her watchfulness, for she had not seen him come after all.

"Only one, Gladys?" she asked, recovering herself.

"Only one that you will care about," answered her sister. "These others seem to be in ladies' handwriting."

Gladys had all the tactless honesty of fifteen years.

Jessica glanced quickly at her sister, then drew her brows together as if pained. Gladys was too gentle to be satirical and too sweet to be unkind. Jessica seemed for the first time to be obtaining an outside view of herself to-day. Hitherto she had been an unconscious Egoist.

"Come here, dear," she said, holding out her hand to Gladys, with a little catch in her voice. Jessica was usually so undemonstrative that Gladys half shyly put her hand in her sister's.

"This letter is from my old German teacher, who is so poor and alone in the world that I am taking care of her out of my allowance. This one is from Mrs. Lloyd Stanley, the mother of Georgia Stanley, and my very dearest friend. She is the cleverest, most human woman I ever knew. She is old enough to be my mother, yet there is more of the spice of life, more of the fun of perennial youth in her, than in any ten girls I know. I would much rather have either of these letters than this one, which is from Frank Fair. Without reading them, I know how they will affect me. Don't think meanly of me, Gladys."

"Indeed I don't. I think the world of you, Jessica."

Jessica pressed the girl's hand in sudden gratitude. Gladys, feeling herself dismissed, left the room with a sigh, wondering for the thousandth time if, when she became a young lady, she would be half as fascinating and successful as Jessica.

Left alone, Jessica read the letters from the two women first. With a little secret thrill of self-contempt, she realized that naturally she would have read Frank's first, if it had not been for what she had just said to Gladys.

When she came to his, she sat turning it over in her hands and wondering what he could have to say. Somehow she dreaded to read it. She had not quite liked the way he had taken his dismissal. He seemed a little revengeful—more mortified than hurt; as if his vanity had been more deeply wounded than his love. He had acted as if his turn would come later.

But this idea of retaliation seemed so impossible, so unworthy of him to plan it, and of her to suspect it, that she put it from her mind. What if he had written to reopen the painful subject? A doubt assailed her, as once or twice it had done before, as to her wisdom in dismissing him. He came dangerously near being what she always had said she must have in the man she married. She had not given him credit for persistence enough to brave a second refusal. If he had — she almost believed she — but no. She had to admit, however, that in the event that he had the perseverance to renew his proposals, it would remove one cause of reproach against him, and she might find herself confronted with a problem.

Then she opened the letter.

A STUDY IN HEARTS

II

With a devout thankfulness that no one can read the egotism of our secret thoughts, she read it a second time all the way through. It was only a note asking permission to present a friend to her, a certain Ramsay Van Buren.

"Where have I heard that name before?" thought Jessica. "Not from Frank Fair, surely. I do believe I have heard Mrs. Stanley speak of him."

She went to her desk and wrote:

"My dearest Friend,—Your delicious letter has this moment arrived, together with one from Frank Fair, saying that Mr. Ramsay Van Buren will be here this week, and that, feeling sure I would enjoy meeting such a man, he had taken the liberty of giving him a letter to me. Is this Mr. Van Buren your Mr. Van Buren? Tell me instantly all you know about him. Isn't he the one who is engaged to Polly Pope? If so he is to be avoided as dangerous. I have heard too much of his fascinations to trust myself with him. I think I shall take to the woods and fast and pray until he leaves town.

"*Why* should Frank want me to meet this man so particularly, unless—

"Hurry and write to me. I will answer your dear letter properly this afternoon.

"Yours with love,

"JESSICA."

Two days later she received Mrs. Stanley's reply, with a special-delivery stamp on it. Jessica laughed when she saw it. But she cut it open with unusual haste and read:

"DEAREST GIRL,— The plot thickens. Your Mr. Van Buren is, indeed, my Mr. Van Buren — alas, not Polly Pope's Mr. Van Buren any more. My dear, that engagement is off, and silence reigns supreme. Of course he does not admit that he broke it—I'd have Lloyd thrash him if he did— not that it is any of *my* business whom he flirts with, to be sure!—but all his friends think he did, so he openly gets the credit for it.

"He is undoubtedly the most attractive man I ever met, and the keenest. *You* never knew anybody like him. Even I, old married woman that I am, with Georgie on my hands to make me ashamed of saying it—even I come under his spell, and feel, when he looks at me, as if he could see the pattern of the wall-paper behind my head. His eyes bore holes in my brain. I feel as if I were being hypnotized in a mild degree. He has a high-bred air and a lofty bearing, "as if he mocked himself and scorned his spirit which could be moved" to play a part upon this insignificant stage of life. He is very conceited. I have heard it said (I feel culpable in repeating to you the gossip of a chit like Georgie and the girls, but you wanted it, so pray forgive me) that he has said that he knew perfectly well that he could have any girl in town for the asking, but that he didn't want any of them. Jessica, I can hardly believe that he said such a dreadful thing. But if he did—flirt with him, Jessie, darling. Bring him to his knees. Make him propose to you, and then throw him over for poor Polly Pope's sake. More girls than she would present you with a wreath of immortelles if you would. You are the only girl in the world who could do it and keep her head, besides being the only one to whom I would suggest such a thing. Oh, I hope the Lord won't punish me for this by allowing some old woman who ought to know better to urge a similar thing on my Georgie! What a weak thing I am! Perhaps he never said it, Jessica.

"I think when you meet him you will agree with me in saying that he is too good not to be better. He needs to have a great deal taken out of him before he can be satisfactory. But he is too clever to be wasted.

"I am quite sure that you will get at the true inwardness of him after all this. Make him propose to you before you get through with him, and I'll give you a dollar! It won't hurt him in the least. It will be good for him. He needs the discipline of a Circe, and that's you, dear. Oh, what a wretch I am!

"I am going to buy my first pair of glasses

to-morrow. I am sick with the thought of the nasty things, but I can't see any more. I may write better after this. So might it be!

"Georgie has gained twelve pounds with the sea-bathing. She would be vexed if she knew I told you.

"Good-bye, dear one.

"Your loving affinity,·

"LUCIA BURKE STANLEY.

"P. S.—I'm sure I don't know why Frank wants him to know you, unless— A man wouldn't be such a fool as to introduce Ramsay Van Buren to a woman he loved, even if the woman *had* jilted him, unless— Oh, I *quite* agree with you! L. B. S."

Jessica felt that this was a matter which deserved thought. She smiled a little derisively at the idea of Polly Pope, or any other of the girls he was thought to have flouted, being grateful to her if she should succeed where they had failed.

"They would hate me for it, no matter how much they might have urged it upon me beforehand. Girls always do. And they would only give me a wreath of immortelles from gratitude if I would be so good as to die and let them hang it on my simple headstone. Thank fortune I don't know Polly Pope or any of them. I wonder when he will come?"

He came that day. He was waiting for her in the drawing-room, and thinking with no little amusement of his lunch at the club with two men who knew her and who admired her so entirely. One had evidently been in love with her and had got over it. He couldn't quite make out whether the other one was now interested and was silent from that cause, or whether he was not yet entangled. He thought it absurd to find one girl turning the heads of so many men—really fine fellows, too. Probably she had mastered a few of the first principles of her art, and these had carried her triumphantly thus far. He hated a flirt. He knew he wouldn't like her. He told himself that he had only called because he knew she must know of Frank Fair's letter. He smiled again when he remembered Woodford's tone to-day when he discovered his (Van Buren's) destination. "Going to call on Miss Drew? The Lord be with you!"

Van Buren was a little disconcerted when he suddenly became aware that Miss Drew was standing before him with a faint smile on her lips at his evident enjoyment of some recollection.

She held out her hand, and he had to cross the room to take it. There was something unusual both in her touch and the way she shook hands. She did not shake hands at all, he remembered afterwards. She put her hand into his and he had held it for a moment. It was different, but it was very pleasant. He was glad he had come.

They covered the first few minutes very well, each studying the other under the cover of the mawkish platitudes by which people begin to know each other. Each discovered that the other was in no way like the preconceived idea.

Jessica was saying over and over to herself: "He never said it. I don't believe it. I can't."

He was indeed unlike any one she had

ever known. He was a power. She knew at once why Mrs. Stanley had said he seemed half scornful of society at large. Mrs. Stanley was magnetic enough to feel this self-appreciation, but Jessica was surprised that her friend's astuteness should have been so much at fault as to name it conceit. Here, at least, was a foeman worthy of her steel. Here was a man to whom she could talk. She began to feel a tingling in her brain, as if her ideas were being called for by one who deserved them, and as if they were waking out of the sleep into which they had been lulled by the conversation of other men.

Van Buren was thinking: "Her eyes are wonderful. They were yellow when I first saw them, and as clear as amber. Now they are troubled with shadows, which come and go. She is worse than beautiful. I wonder what she is thinking about? I wonder what she thinks of *me?*"

This question was a new one. There never before had been any doubt in his mind as to what any girl thought of him. Most girls were transparent, and he, unlike most men, was brilliantly interested in the study of human nature.

Jessica recognized his interest.

"He is regarding me as a sentient being," she said to herself. "Perhaps he suspects that I am capable of thought. Heavens! he must not know that! I must be more frivolous!"

Her curiosity to know just why Frank Fair had sent Mr. Van Buren to see her was tempered with a certain fear of the truth when it should be discovered. He had come so near being the right one for her, yet had just missed it by a hair's-breadth. She had not quite got her own consent to trust him. Her faith was not intuitive. She had reasoned herself into it because she felt ashamed of being so ignoble as to refuse it. Her present predicament was precisely the one she had been in many times before. Instead of being able to credit him with the noblest qualities, she doubted him persistently. She felt herself grow hot with self-indignation that she stooped to credit him with vengeance so unworthy as to introduce to her a man so well known to be attractive as to be dangerous. Jessica was so hopelessly honest with herself that she tried to believe it was only a base reflection of her own base suggestion to avenge poor Polly Pope.

"Evidently you will not feel at ease until you have decided," said Mr. Van Buren.

Jessica looked up so plainly startled that he smiled.

"How did you know?" she said, bravely.

"How fine of her not to stoop to equivocate!" he said to himself.

"Oh, by the lights and shadows in your eyes. Your mind is wrestling with a problem. I cannot flatter myself that I am the problem, but you are wondering if I can help you to solve it."

"I wish there were more men like you," she said, simply. "It is gratifying to meet one who expects a woman to think."

Van Buren made a mental recoil. She had done precisely what he had been waiting for. It was an old trick. It was one of his own, therefore he recognized it. She had struck the personal note. All coquettes know how to do it. Only few do it with quite the air of simplicity that Jessica used. Her voice sounded grateful. Nevertheless he determined not to fall a victim. Her fascinations were too well known.

"Then I was right?" he hazarded.

"You were quite right. I was wondering if a woman ever were justified in disregarding her intuitions."

"I don't know. We men have none."

"I know you are said not to have, but that reason or logic which gives you a hold over another human intelligence, such as you have just displayed, must be akin to intuition."

Van Buren felt such a satisfaction permeate his being at the subtlety of her remark that he pulled himself up with a start. He was disgusted to find that she was actually making headway with him after all his warnings. He was annoyed, too, to find that her tones had such sincerity in them—that her flattery sounded so much like the simple truth uttered with a child's directness. This, then, was her game, he thought. She steals "the livery of the court of Heaven to serve the devil in!"

"Men, I believe, possess but five senses, women six, and coquettes seven," he said, looking her directly in the eyes.

Jessica's electrical nature felt the sudden change, and she knew just how it had come about. She had seen other men grope blindly through it. This one was clever enough to extricate himself. Nevertheless, she felt the blow. As she did not reply at once, he added:

"That is where a woman has the advantage over a man."

His ironical tone stung the girl.

"And a brute over both," she said, with her amber eyes turning dark, and the scarlet coming into her lips. "In the brute it is instinct; in woman, intuition. Man is the only creature sent helpless into the world, to blunder along on reason."

Her recognition of his perception and her swift reproof he regarded with an admiration he never yet had accorded to any woman. His sense of humor and his appreciation of her pluck overbalanced his chagrin, and he laughed.

Jessica only smiled. She was sincerely

hurt. Most men believed her to be better than she really was. Here was a man who would not believe her to be even as good as she knew herself to be. She was stirred beyond all reason. The placidity of both was ruffled. The positive and negative had come together with a flash which was at once disquieting and exciting.

"You believe, then, in instinct over intuition?" he asked.

"Brutes follow their instinct blindly if left to themselves, and never make mistakes unless interfered with by human beings. Women only make mistakes when they disregard their intuition. Men, pertinaciously following reason to a logical conclusion, make the most mistakes of all," she answered.

There was no bitterness in her tone, only deep feeling.

"How so?" he asked.

"By satisfying their own intelligence regardless of the fact that in so doing they often wound another's heart."

Van Buren regarded her suspiciously to see if she were becoming personal. But her cool tones and clear eyes assured him that she was but generalizing.

He was surprised to find himself disappointed that she was not personal. He resented this weakness in himself to the extent that he rose to go.

"A clever woman like yourself need never fear that her heart will ever be wounded by anything," he said, unwisely.

Jessica was as quick to take advantage of the opening as if she had read his look.

"Were you personal? I am afraid I was only generalizing."

He bit his lip in momentary vexation. Then something in the slightness of the girl as she stood before him, something spiritual in her expression as she raised her eyes to his, appealed to him more strongly than anything she had said or done. He felt that as a man he had not been just to her, a woman. They had come into each other's power only to feel that they had begun wrongly, and had placed each other at a disadvantage. Van Buren's fine generosity made him put something of this into words.

"I wish we might begin over," he said, simply. "We could do better."

Jessica recognized instantly what he had left unsaid.

"No," she said; "let us go on from now. Only—" she hesitated and put her hand in his—"only do not believe me to be so clever that I am not a woman, with all a woman's helplessness."

III

Van Buren turned away from the touch of Jessica's fingers, and the look on her face more nearly conquered, in spite of all his warnings, than he would have believed possible. Her personality seemed to tingle in the innermost recesses of his being. She touched a chord he did not know he possessed. She appealed to him. Great Heavens! how it appealed to a man sated with commonplace flattery to have so clever a woman lay down her arms at the outset and beg a truce—a truce not so much for her own protection as because she knew it was wasting the time of both! His imagination, stirred by her strange beauty, supplied all that she had not said. She meant him to understand that fencing would be so futile, so unworthy. Why should they not mutually capitulate and continue to explore?

There was nothing of the coquette in her appearance. The delicacy of her womanliness, the spirituality of her mental vision refuted the idea of the coarseness of feeling which the word implied. She had stirred impulses in him which no one had ever called out before. She had appealed to the higher nature he always had suspected himself of possessing but never had been able to prove. And now that she had proved it so unwittingly and so unconsciously by just permitting him to catch a glimpse of her sweet woman's soul, gratitude alone would urge him to respond—gratitude and— Hang it all!—*was* it gratitude? Wasn't it gratified vanity? The devil take it!—wasn't it *love?*

He stopped short and struck his stick so sharply upon the pavement that two men in front of him looked around.

"Pay no attention to me, gentlemen," he said to himself. "I am not even a lunatic. I am only an ass."

IV

Left alone, Jessica fled to her room and closed the door.

"It's come!" she said to herself, pacing up and down with her hands wrenched together. "It's come at last—the retribution I've been waiting for. I knew it would overtake me some time. I've warned myself too many times to shrink now. Oh, Frank was clever!—Frank was very clever! He knew me better than I thought. He has avenged himself already. He has made me meet my Fate, knowing perfectly well that I never could penetrate his armor. Now it is my turn. Oh, I knew it would come! but that does not make it any the less hard. I ought to hate Frank for it, but I don't. I am glad of it. I am glad to have known this man. Now that does not mean that I have any hope—yes, it does, too! It means that I think I shall see him again, although he did not say so. It means that I think if I have been successful with so many, many others, why not with this one? Oh, but he is not like the others! He is not like any one I ever knew before. That is just what Mrs. Stanley said. She warned me. Everybody warned me. Yet here I go, blind and deaf, insisting on my own discomfiture, fairly overjoyed at the idea that I am going to be made a fool of, and determined not to draw back even while I can. While I *can*, did I say? Why, is it not too late already? Could I ever forget him even now? From this time on must he not be a part of me, only—I must remind myself—*only* to end in sorrow and distress for me. *He* will not care. *He* is an older hand at it than I am. If I were not so despicable, if I had lived my life as nobly as I knew how, if I had followed the highest, I should at least not be tortured by these doubts and apprehensions. I suppose women who never have encouraged a love they did not intend to return never dream that an honest love might not be reciprocated.

They are not tortured with remorse. They do not dread that the fair appearance on the man's part may be only the cloak of trifling of which they so often have made use themselves. What a fool I am to think that he will care for me! Look at poor Polly Pope and all the others! I am only the last one. In three months from now people will be saying, 'There was Polly Pope and poor Jessica Drew!' Oh, I could not bear that! I could not!—I could not! How noble I am! I believe I dread what people would say more than any pain I might suffer! If I might only suffer in private and have no one know! Oh, I am a magnificent creature! *Any* man ought to be proud of my love—it is so worthy, so lofty!"

She knelt down in front of a chair and laid her head on her folded arms.

"Yet," she whispered, "it's *in* me to be noble, if only I had half a chance. If I loved a truly good man, I could forget everything and be different."

She sat there thinking for some time. It grew dusk. She sprang up suddenly and found Mrs. Stanley's letter. She read it over deliberately twice, then put it back in the envelope.

"That settles it," she said, aloud. "I will at least have the courage to save myself, now that I can reason it all out. Mrs. Stanley is at Narragansett Pier. I'll surprise her by accepting her invitation. I'll go to-morrow."

V

Jessica colored under her veil as she settled herself in the train the next day, to realize that she was running away—she, Jessica Drew, actually running away. She did not feel as comfortable as one should when she is doing her duty and saving herself from a palpable danger. Instead, she felt miserably uncertain and reluctant to go. She felt that she had been foolish to be so precipitate. She ought to have waited to see him just once more. She smiled a little bitterly to realize that she was only cheating herself, and that the reason she felt that she ought to have seen him again was just because she wanted to see him—yes, she admitted it freely to herself—she wanted to see him again, more than anything else in the world.

As the train rushed on, some one entered from the smoking-car. She looked up and encountered Mr. Van Buren's startled gaze.

He hesitated a moment palpably, then came and sat down beside her, placing her wrap on the front seat.

"That seat was mine," he said, with a curious, repressed smile of self-derision.

"How strange," murmured Jessica, "that I should have taken this one."

"Not at all," he answered, grimly.

Jessica looked up at him timidly. He had not once looked towards her.

"I thought you meant to stay in town longer," she said, at length.

"I did. I came away very suddenly. I have just written you a note and sent you some roses."

He seemed to hurl these admissions at her. Jessica felt a wild impulse to telegraph that the note and roses be forwarded to her. She could not bear to miss the first flowers he had sent to her—in all probability the only flowers he would ever send.

"Oh, thank you so much," she said. "I—"

"You did not tell me yesterday that you intended to leave town to-day," he interrupted her.

"Did I not?" she said, carelessly, regaining her hold upon herself as she saw him losing his. Inside her frock was Mrs. Stanley's letter, which she had brought to brace herself with if she found herself weakening. She placed her hand over it and heard it crackle. She wondered how far he was going. She felt her resolution weakening. It was so *good* to be near him again. He was such a power.

"I am going to Narragansett Pier," he said, presently. "I suppose you are going there also?"

Jessica caught her breath in sudden terror.

"No," she said, hurriedly, "I am going to"— where in Heaven's name was her Aunt Livingstone's cottage?—"I am going to York Harbor."

"But this is not the train for York Harbor," he said, looking at her for the first time, his man's mind driving out all thought of love at the idea of a woman's having got on the wrong train.

"Oh!" said Jessica, coloring hotly. "Can my father have made a mistake? I thought I told him I wanted to go to York first."

"Then you meant to go to Narragansett later?"

"Yes. Mrs. Stanley is there."

"Then go to her first," he said, eagerly.

"No, no," cried Jessica. "I must go to York Harbor. You must arrange it for me. You must help me. My aunt is expecting me."

Her eyes fell as she remembered that her aunt at that very moment was dangerously ill, and that she could in nowise be received there.

"Give yourself no uneasiness, I beg of you, Miss Drew. I will take you there myself," he said.

Jessica laughed a little hysterically.

"You must do no such thing," she said. "I will go on to Narragansett."

Van Buren was chagrined to realize what a sweep of joy those words gave him. He looked down at her flower-like face and deliberately gave himself up to the current of his passion.

VI

It was the curse of triflers such as these that, in the month which followed, neither Jessica nor Van Buren could take all the comfort from the blissful pastime of falling in love with which that rapturous occupation is usually fraught. Each was conscious of his own sincerity for the first time. Each constantly doubted the other's. Mrs. Stanley was overjoyed beyond words to express to have so delicious a flirtation brought before her very eyes, and unconsciously she augmented Jessica's distrust by her laughter, warnings, hints, advice, all put in so clever a way that Jessica would rather have died than admit the truth.

"Jessica, darling," Mrs. Stanley often said, "it is as good as a play. I never saw such acting. I never should dream that he was not desperately in love with you if he had not treated poor Polly Pope in such a manner. I believe he is capable of anything. Go on with it though, dearest. You can bring him to his knees if any one can. He will be a different creature."

"I can't bear to hear you talk that way, Lucia," said Jessica. "Suppose we are wrong about Polly Pope?"

"I know it, my dear. It is wicked of us. Oh, *what* if some one should ever urge my Georgie to flirt as I have urged you? Suppose we stop it at once and behave ourselves? No harm has been done as yet."

"Of course no harm has been done as yet!" echoed Jessica, drearily.

"You are not natural with him, Jessica, darling. You sound artificial to me, who know you so well. He must be clever enough to detect it."

"I dare not be natural with him. I am on my guard constantly, lest I be caught napping."

"It must be a great bore," observed Mrs. Stanley.

They all went back to town together.

To Van Buren it seemed that he never could leave Jessica voluntarily. If he could have believed the evidence of his own senses, he would have ended his suspense long ago, but the words of his friend, Frank Fair, kept running in his mind with fatal persistency. "Remember, old man," he had said, "she is the most consummate actress you ever met. Everything comes from her head. I never yet have detected an act which came from her heart—if, indeed, she has one."

These words told more to a man like Ramsay Van Buren than Frank Fair dreamed, and it was on account of them that Van Buren remained in New York, leaving Jessica to return to her home without telling her that he loved her.

The autumn passed away. Van Buren heard of her at Lenox just in time to escape going there himself. He purposely avoided Frank Fair, dreading to hear any more on a subject which was rapidly becoming the main issue with him.

The day before Christmas he was walking down Broadway, thinking how absurd it was to let any one influence him in a matter which Jessica alone could settle. She had come to be so close to him, so high and fine since he had left her, that he no longer thought of anything except that he wanted her; he wanted her in his life more than he wanted anything else in the world.

He suddenly decided to go to see her—to go that very day. A weight seemed lifted from his heart the moment he did so. He paused on the corner of Twenty-third Street to light his cigarette. A chill, eager wind swept up the street and blew out the match. He pulled up his fur collar and tried again. Again the match went out. He tried a third time with the same result. With an exclamation of impatience he flung down his last match and moved on, when a little newsboy, with a shrewd, merry face, lighted a match which burned bravely in spite of the wind and held it up, hopping backwards in front of him and saying:

"Please, mister, lemme light yer cigarette for yer! Please, mister, lemme light yer cigarette for yer!"

The boy looked as if ready to dodge a

blow, but he persisted in his attentions, until Van Buren, with a laugh, stooped and took the match. As he did so, another and daintier laugh fell upon his ear, and there stood Jessica Drew beside him, the snowflakes standing out in crystals upon her velvet and fur coat.

"You here!" he exclaimed, in delight. "I was just going to run down to see you."

"Were you? Well, I have saved you the trouble, for I am here. I came up to spend Christmas with Mrs. Stanley. She goes to California on the 3d."

He turned to walk along with her, his eyes devouring the lovely tints of her face and hair, not noticing that the little newsboy was still hopping along backwards in front of them, smoking the cigarette he had snatched ere it reached the sidewalk, his shrewd blue eyes glued to their rapt faces.

Jessica was the first to see him.

"Why, look at the child!" she exclaimed. "His shoes are so worn his feet are almost upon the ground!"

"And to-morrer's Christmas, loidy," he flashed, with the quickness of the gamin.

"So it is, my boy," said Van Buren. "Take this and buy yourself some shoes."

"Not on yer life, mister. I'll take it home to me mudder, if yer don't mind."

"The dear child," murmured Jessica.

"The little rascal," murmured Van Buren.

"Is it a trick?" she whispered.

"Nearly everything is," he answered.

She bent over the boy and spoke to him.

"Mrs. Stanley's carriage is waiting for me," she said. "There, I see it! Can you attract Summers's attention? Thank you so much."

"Shall you be at home this afternoon? No? To-morrow? May I come for an hour on Christmas Day?"

"Yes, do. We shall be so glad to see you."

Something in Jessica's eyes went to his heart. When he had seen her drive away, Van Buren hailed a cab and collared the newsboy.

"Come here, you young rascal!" he cried, in an exuberance of joy which none of his friends would have recognized. "Get in here with your papers and your everlasting matches, and your tricks and your knowledge of humanity, and let us make a call upon that mother of yours! I suppose she is sick? No? She wants coal, doesn't she? I thought so. I suppose she has nine children, all smaller than you? No? Only three? How moderate for a poor woman, who might have so many more! I suppose your father drinks? No? Dead, is he? I am surprised at him. He ought to be alive and drinking. Are you going to have a Christmas-tree to-morrow? No? How dare you lie to me like that! I say you *are* going to have one! And a Christmas-dinner besides, with turkey and oysters and ice-cream, and I want you to eat until you burst, do you hear me? Will you promise to burst?"

The boy roared out with delight.

"I suppose you are pleased, you young imp, to think how you are taking me in. I don't believe you have any mother or three small sisters. I believe your mother is wealthy and that your father belongs to the Church. I know you are chuckling to think what a snap you have; but, do you know, I don't in the least care whether you are lying or not? Do you know, I am bound to give coal and a Christmas-tree and a Christmas-dinner to somebody, and that you are going to be that somebody? And do you know why?"

"It's all along of der loidy," said the boy.

"Guessed it the first time," said Van Buren. "Go up head. 'It's all along of der loidy.' And if I ever do another thing of this kind, which might well alarm my friends for my sanity, it will be 'all along of der loidy.' If I ever amount to anything in this world, boy, or if I go to the devil, it will be 'all along of der loidy.'"

"Here we are, sir."

"Lord, what a place!" cried Van Buren, as he followed the boy into a dark doorway.

VII

On Christmas Day, too early in the morning for Ramsay Van Buren to dream of meeting any of his friends, he again betook himself to the same dark doorway, and amid the clamoring of the three children, whom the boy Tom had not lied about, he began to set up the most astonishing of Christmas-trees.

"I don't exactly know what is the matter with me, Mrs. O'Reilly, but I really feel as if I could help you to order your Christmas-dinner. I am very fond of cake with my ice-cream. Have you any cake, Mrs. O'Reilly? No? Then will you kindly step around to the bakery and buy one, with a great deal of icing on top, and—and anything else you may see, Mrs. O'Reilly. And if you would be kind enough to take the children with you and buy them some candy—buy them plenty, to make themselves ill, Mrs. O'Reilly; it will not be a successful Christmas without getting ill—it will save the baby's having to eat those raw cranberries, which she is now proceeding to do; besides giving me time to get these presents on the tree," he added to himself, as the four went out, laughing and calling down Irish blessings on his head.

"Where in the deuce shall I put these things?" he said, helplessly, holding some large red popcorn-balls up by a suspiciously slender string.

A slight rustling in the doorway made him say:

"I feel that I must have some help, Mrs. O'Reilly—"

"I am not Mrs. O'Reilly," said a voice, "but—"

He turned suddenly, and saw Jessica's beautiful face over the top of her bundles.

"Jessica!" he exclaimed, dropping his sticky burden and taking her bundles from her, with a face so full of delight in her presence that she colored beautifully, and under his gaze the tears gathered in her eyes. She looked up at him.

"Forgive me," she murmured. "I have been so unjust to you."

He held out his arms to her.

"Will you love me if I forgive you?"

"I love you now," she said, simply. "I have loved you for a long time."

He drew in his breath sharply at the royal simplicity of her admission. He had expected to be obliged to use every art to coax the assurance from her which her eyes had given him the day before. He had expected—Heaven only knows *what* he had expected.

"You have loved *me*, Jessica? Oh, why didn't I know it? You have made me so unhappy, Jessica. Why didn't you tell me?"

"Why, how could I tell *you?* You never asked me. You never trusted me. I never could be myself before you. You prevented my coming here when I left you, for fear you would think my charity but another wile. I have hardly dared to be as good as I wanted to be since I knew you!"

"Oh, Jessica, how charming you are! But how much time we have wasted!"

"We have not wasted a moment. We never could have arrived at this stage if we had not been looking at ourselves through each other's eyes. Then, too, I never dared let myself go after what I heard of poor Polly Pope."

"'Poor Polly Pope' happens to be engaged to an imbecile millionnaire just at present. Did you believe that I was engaged to her? I never was. She could have stopped the talk in a moment if she only would. *I* could say nothing."

"I knew you could explain it!" she cried. "I felt that most positively when I saw you yesterday."

"Was that why you looked at me when I closed the carriage door?"

"Did I look at you?" murmured Jessica.

"*Did* you look at me? I wonder if a woman ever looks at a man like that without being conscious of it? So you have believed in me since yesterday?"

"And that is longer than you have believed in me, by twenty-four hours."

"I told you yesterday when I saw you that I had just decided to run down to see you. Of course you knew what that meant."

"It meant that you had determined to allow yourself to love me in spite of all you had heard—in spite of everything. You did not really believe in me until just now —until I told you that I—"

"That you *loved* me! Oh, Jessica, *dear* Jessica!"

"It's true! That is what I told you.

How helpless a man is when he depends on reason! I had *faith* in you."

"No, no, dear love. I was afraid of you, that was all. I had heard so much about poor Frank Fair and the others. All the fellows who hadn't been in love with you were just going to be. Everybody warned me. I never dared hope that you could care for me. I don't deserve you, Jessica. I can only say that I have loved you from the first time I ever saw you, more and more every minute and every hour, until even you, who have been loved by so many, should be satisfied."

"Don't talk about the others," she murmured. "They have made us trouble enough. Think only that you love me and that I love you!"

"I got onions, Misther Van Buren," cried a voice on the stairs, "thinkin' you might enjoy the taste of an onion this blessed Christmas Day."

"Thank you, Mrs. O'Reilly, it isn't exactly onions that my mind is fixed on to-day. This is Miss Drew—"

"Bless her dear face, she was here yesterday after you left! I just know her to be God's own angel."

"Indeed, Mrs. O'Reilly, you are a woman of wonderful discernment. Where is Tommy?"

"Tommy is sick, sor, wid the candy."

"Good. I am glad to hear it. That makes Christmas quite complete — that, and what has just happened to me. As soon as you feel better, Tommy, eat some more and get sick again. If you have to have a doctor, I'll pay for him most cheerfully. I couldn't invest my money better."

"Well, saints preserve us," said Mrs. O'Reilly, when Jessica and Van Buren were gone, "did you ever see such a free-handed gintlemin? Offering to pay the doctor so handsome. What a pity one of us couldn't be took sick to-day instead of last week! Ah, he is a foine man."

"It's all along of der loidy," said Tommy, uncoiling himself in his corner.

Mrs. O'Reilly hung out of her window.

"They come in two carriages," she said, "but they wint away in wan."

THE HEART OF BRIER ROSE

THE HEART OF BRIER ROSE

I

THE Weeping Willow telegraph-office faced the level prairie. Up and down before it, like shining ribbons, lay the railroad tracks, converging mysteriously until distance blended them into one. Back of it flared the wide main street, with stores and cottages indiscriminately mingled, which marks the disconsolate prairie town. Beyond, enclosed by a white picket-fence, straggled the desolate graveyard. Some sorry-looking brutes, with down-hung heads and burrs in their tails, were hitched in front of the post-office. For the rest, there was the vastness and lonely boundlessness of the never-ending prairie. Mounted guard over this living finger-post, quivered the remote sky, with its unblinking Cyclops eye. It was a barren and meagre settlement of families from other states, casting their lots together to make a scanty whole, and forecasting their dreary life's end by naming their village Weeping Willow. The only thing in plenty which Nature supplied was room. There was an abundance of space. It was quite a walk to cross the street. Neighbors' houses stood aloof. Nobody was crowded, even in the graveyard.

The telegraph-operator, satiated with landscape, leaned back, stretched himself prodigiously, yawned audibly, and collapsed in his chair, which creaked in vexed remonstrance. He tossed a remark over his shoulder. "So this is what you are yearnin' fur, Dave?"

Dave took his cane, and, limping to the door, viewed the inertness in silence. Then he roused himself, and said, cheerfully:

"A telegraph-operator is all I'm good fur since I got hurt. Thankful enough I'll be if I get the Red Valley job. I'd like to be so near to you, Joe."

"Seems like the comp'ny might have done

more for you when you got smashed up in their own accident. 'Twouldn't have hurt 'em none to keep you as conductor," grumbled his friend.

"How could a lame man be a conductor?" returned Dave, with his unfailing good-humor.

"A railroad company is a measly concern on the pay act," observed Joe, gloomily.

Suddenly the afternoon stillness was broken by excited voices, and the sharp barking and yapping of dogs. Joe brought his feet to the floor in a hurry.

"I can't leave the machine, Dave. Go and see what the rumpus is about. I'll bet Brier Rose is up to somethin'. It takes that there girl to stir up the boys. No, Foxy," he said to his terrier, who was whirling around in an ecstasy of anticipation, "you stay here. If Brier Rose *is* at the bottom of it, a little feller like you might get lost in the shuffle."

Dave obediently limped up the street, where, in the midst of a crowd of rough men, stood a girl holding some little animal high above her head, while the dogs leaped and snapped around her. The men were laughing and keeping the dogs partly in check. The girl, with scarlet cheeks, begged and scolded and threatened them all, to their infinite amusement.

"Call off your dawg, Jim!" she said, fiercely, to the owner of the largest, whose leaps sometimes almost reached the quivering little object in her hands.

"Throw down the beast an' I will," he answered.

She turned her flashing eyes on him. "If that there dawg gives another jump, I'll pizen him before sun-up," she said, slowly.

Jim made a lunge for the dog, and sat on him to keep him down, while the crowd hooted in derision of his obedience.

"What's all this?" cried Dave, coming up and pushing his way through their midst.

"Brier Rose is bein' held up!" cried a voice.

The crowd yelled with delight. The girl's whole face became white with rage as she singled out the speaker.

"You'll pay for that, Ben Miles, as you've paid before," she said.

THE HEART OF BRIER ROSE

Again they shouted at some recollection. Even Brier Rose condescended to laugh, angry as she was, and Ben subsided.

"Call off those brutes," cried Dave, rapping the nearest dog with his cane. "For shame, to tease a woman!"

"Look a-hyer, stranger," said a young giant, menacingly. He towered above Dave, who stood his ground.

"I'm lame, and no account in a fight," said Dave, "but half a man ain't goin' to see a woman tormented."

"Who in thunder—" began his threatener, but Ben Miles laid a hand on his arm.

"Hold on, Jim," he said; "that there's Dave Comstock, conductor of the smashed-up No. 7."

"Not the feller that got hurt savin' the baby?"

"The same."

"Sho, stranger!" said the mollified Jim. "You're welcome to interfere. Give us yer hand. We wouldn't hurt her fur nothin'. Bless my stars! Brier Rose can take care of herself better'n most men."

"You see, it's seldom we git a chance to

git even," explained Ben Miles, as the men closed around Dave to shake hands.

"Get even fur what?"

"Fur what? Lord, stranger, you must come from parts unknown! Fur everything! Ther' ain't a man in Weepin' Willer but what's been sassed by Brier Rose. Sassed? Lord! I should say so. But we wouldn't hurt her none. Stranger, lemme give you an *ad*vise: don't you worry none about Brier Rose!"

The crowd roared. The dogs were all held now, and the girl put her tired arms down. She looked curiously at this man, whose brave story she knew by heart, as she heard him defend her. To be sure, she had been defended before; there was hardly a man who would not have risked his life to save hers, but they teased her unmercifully whenever they got the chance. Dave's interference was on a new line. She did not quite understand it, but it appealed to her at once. She wrapped her apron around the little animal, and with a new sensation stirring at her heart Brier Rose slipped away.

THE HEART OF BRIER ROSE

When Dave went back to the station to tell Joe, the latter roared with delight.

"Just like her! Edzackly like her!" he cried, slapping his leg so inhumanly that his lame friend winced for him.

"Who is Brier Rose?" he repeated, in answer to Dave's question. "You don't know much if you don't know old Bryan's daughter. She's the best-known girl from Horseshoe Gap to Powder Crik. Old Bryan's been engineer on the road ever since the track was laid. There's them that can remember his takin' the child with him on the engine when she was a little mite of a thing. All eyes she was then, as she is now. What wasn't eyes was temper. Same now, savin' that now she bosses the boys in addition to old Bryan. She can run an engine with the best of 'em. Bryan's taught her all the tricks, and he thinks the sun rises and sets for just her. Sharp as chain-lightnin', is Brier Rose, an' prickly as a cactus. All the boys been touched in the head about her at one time or another, but she sasses the hull of 'em. It's my belief that she'll wipe the earth with Ben Miles for that there 'Brier

Rose is bein' held up.' He'll wisht he was dead before she gits through with him. You hear me!" And Joe's noisy mirth made the windows rattle.

"Strange she would defend a gopher, when she's so hard on the boys," observed Dave.

"That's just it! That's Brier Rose! She's got more tame pets! She's friendlier with every beast in Weepin' Willer than with any of the boys. She's just that curious. She ain't even got a head fur anybody but old Bryan—you notice I make no mention of heart concernin' Brier Rose; I don't keer to talk of what she ain't got— and just now she's specially bewitched about him. After keepin' straight for forty years, he's taken to drink. The girl knows he'll lose his job if the company gets wind of it, and she watches him like a hawk. Many's the time she's made his night run with him, for fear he'd lose his head. It's only at times he goes on a tear, and she knows the signs. Brier Rose is proud as Lucy Fire, and knows how to keep her mouth shut (which the same I can't say for most women;

wisht I could); but we all know about it, and look out for her."

"What's Bryan's run?"

"Horseshoe to Powder Crik. She knows every inch of track and siding. And I wisht you could see her handle the critter. She knows all Bryan does, and she's a heap sight quicker calc'latin' than the old man. She can tell about how fast any train's goin' if it just runs by her, or if she's on it—either one. It's wuth while to see her oil and clean the machine. She goes over it spry as a kitten."

"She's handsome," said Dave, simply.

"Humph! Handsome is as handsome does," observed Joe, grumpily. "She's cold as ice and hard as a rock. It's my belief that she 'ain't got no heart same as other wimmin. And as to love! Whew! Stand off! There's a touch-me-not for you! And sassy? Lord!"

"*All* the boys been touched in the head about her, I believe you said," remarked Dave, mischievously.

Joe hitched at his left suspender uncomfortably and slowly reddened. "Nigh about—nigh about," he said, hastily.

THE HEART OF BRIER ROSE

Dave looked down the glistening tracks, which seemed to stretch imploringly into the future. "I hope I'll get the Red Valley job," he said, abruptly.

In spite of what he had heard—or, perhaps, because of what he had heard, for who has language subtle enough to follow the intricate pathway of a human soul?—all things, even the melancholy town itself, grew rose-colored to Dave's sunny eyes. With his usual unfailing cheerfulness, he waited hopefully for news of his appointment at Red Valley, and hovered, as if fascinated, around engine Forty-four.

Neither the boys nor old Bryan were slow to notice this, the latter having accepted such attentions periodically from all the young men. It was so inevitable a proceeding that, up to the time of the Middleton's dance, they paid no attention to it. But that night something extraordinary happened.

The next day, as Brier Rose rode down the street on her hardy Texas pony, the boys gathered around her eagerly, notwithstanding the fact that she had a stout little whip in her hand. They had something new and strange to tease her about.

"Brier Rose," called out Jim, as she drew rein, "you don't care nothin' about dancin', do you?"

"You'd ruther *set* all the evenin', wouldn't you, now?"

"D'you like the name o' Dave, or do you reckon you'd ruther have Comstock?"

Rose looked from one to the other as the bottled-up taunts fell rapidly upon her ears, her cheeks and lips growing scarlet. For once her ready tongue failed her. Small need to ask them what they meant. Too well she knew. But was her subjugation apparent in such a trifle? And so soon? And Dave, as yet, had said nothing. Oh, shame! shame! Her eyes smouldered dangerously, yet with all her gathering fury there was an odd fluttering in her white throat and a cruel pain at her heart. Emboldened by her silence, they went further.

"What does *he* say about it?"

She thought she detected the mockery of the question, being all unaware of Dave's interest in the Forty-four. The shamed crimson leaped to her very temples and receded, leaving her face pitifully white. The boys must know that poor Brier Rose was ready to be plucked before she was sought, or they would not *dare* to speak of it! Her wounded pride now panted for but one thing—a way out. She saw him coming down the street.

"Do you love him? Say, Brier Rose, do you love Dave?" cried the one farthest from her whip.

Her courage came back at Dave's approach, and the spell of her unwonted silence was broken. She laughed scornfully as she saw that he had heard the question and had involuntarily paused for her reply.

"Do I love him?" she cried, looking him fairly in the face. "I come nearer to hatin' him!"

She turned her horse sharply, and the blows the boys had expected fell on her fiery little Texan. He craned his neck and went up the street on a dead run; but fast as Rose flew, the grieved look in Dave Comstock's blue eyes kept pace with her.

* * * * *

That night Joe fidgeted around, unable to decide whether or not he should speak to Dave about the occurrence of the afternoon. Dave's genial smile and cheery hopefulness were gone. He sat with his face buried in his folded arms.

Joe drove his hands deep into his pockets, and strode up and down the room stormily. Finally he burst out with:

"Dave, don't mind it, old feller. I told you she hadn't a heart for a man; she wastes it all on dumb brutes. She ain't worth grievin' after. I'm sorry you ever saw her. You're not? Well, of course, if you feel that-a-way, I've nothin' more to say. Only, Dave, my boy, you're too good for her."

"No, no, Joe, don't say that. We're none of us good enough for a woman when it comes to that. I don't blame her. Of course I'm lame, and you can't expect a woman to love a lame man when there are so many others, can you, Joe?"

Joe coughed noisily, but said nothing. Dave looked down at his poor, maimed foot.

THE HEART OF BRIER ROSE

"Joe, do you know that little baby I saved from the wreck had brown eyes like Brier Rose? I remember the baby smiled when I held it out to the men. You know my foot was caught and I couldn't move. I've never seen Brier Rose smile at me that way. If I had saved *her*, perhaps she would. Do you think so, Joe?"

* * * * *

At home, Rose was thinking of the story of Dave's bravery in the wrecked train, of the lives he had saved, of his defence of her. And to-day, in return, she had mocked him. Ay, if the look he gave her spoke truly, she had cut him to the heart. Tears—tears in the eyes of Brier Rose!

II

The position of telegraph-operator at Red Valley was given to Dave Comstock. The afternoon freight, heavily loaded, had just pulled clumsily out of the Weeping Willow station, with Dave on the rear platform of the way-car. Joe waved his hat to him from the

THE HEART OF BRIER ROSE

window, and Brier Rose, from the cab of the Forty-four, looked wistfully after him, and shook her shining hair over her eyes to hide their dimness.

The Forty-four, having come down on the rear of the freight as second engine, now stood on the siding, waiting to go back to Horseshoe for the midnight express. Old Bryan was up in a crowd of men in front of the post-office. Brier Rose watched him anxiously. As long as he kept away from The Owl she felt easy. He knew she was watching him. He also knew that she would not hesitate to come after him if The Owl proved too strong an attraction. Therefore he kept away.

Joe stood in the doorway, and admired Brier Rose against his will. He still was sore at heart over his friend, and fear of getting the worst of it alone prevented him from giving that girl a piece of his mind.

She trod fearlessly along the side of the boiler, rubbing the hand-rail with a black, oil-sodden cloth. She touched the engine as if she loved it. Every part of it shone like the sun. Her long-pointed oil-can had done its work. Every valve worked with precision. Each screw was secure. Joe laughed to see her fling a shovelful of coal into the furnace like a born fireman.

His own machine called his attention from the Forty-four. Then Rose heard him cry out, and, springing down, she rushed into the station.

"A runaway engine coming this way!" he said, hoarsely. "Spite work of a discharged engineer. No one on her—going twenty-five miles an hour—a single track—Dave's train only going fifteen—the Forty-four and that ore-car on the only siding between here and Red Valley. My God!"

"Where is it?" cried Brier Rose.

"It broke away from Horseshoe Gap. Message is from Prairie City. It's already passed Prairie City, headed straight for here. It's bound to catch Dave before his train gets to Red Valley."

Rose turned white to her very lips. She covered her face with her brown hands. Only for a moment, though. Then she flung back her head, and looked Joe full in the face.

THE HEART OF BRIER ROSE

"I can save him!" she cried. She sprang for her engine and climbed into the cab.

"Rose! Rose!" roared Joe, in dismay.

Rose turned her white face towards him imploringly. "Be at the switch, Joe, and listen for my signals, as you value Dave's life!" she cried. Then she pulled the throttle-valve out to its full extent. The engine shivered all over; and, at fifty miles an hour, the Forty-four, driven by Brier Rose, leaped down the track to meet the runaway.

There was not a moment to lose. A certain number of miles, lessening every moment, lay between the lumbering freight with Dave on board, and the cruel, senseless, runaway engine. Between them was Brier Rose, with just a chance of safety.

Feverishly she examined the familiar machinery. Eagerly she scanned the track for signs of the runaway. She knew that a loosened rail or any obstruction would hurl her to her doom, and still not avert disaster from Dave. The whistle of the Forty-four shrilled out an unearthly screech continually, to warn even the birds from fluttering too near the messenger of life. The prairie-dogs scuttled into their holes in fear. The telegraph-wires intoned. The bending sky took on a new meaning to her. The engine rocked from side to side at the dizzy rate of speed. For the first time the odor of hot oil made Rose feel faint. She hung half out of the cab-window, panting for breath, with her hands clinging crazily to the window for support.

Suddenly she saw smoke in the distance. Larger and larger grew the black speck on the track. Faster and faster flew the Forty-four to meet it. Nearer and nearer came the runaway. When she could plainly see the shape of the approaching engine, she closed the throttle with a rush which made the Forty-four tremble. She reversed her engine, and, at little less than twenty-five miles an hour, began running away from the runaway.

Slowly, almost imperceptibly, it gained on her brave engine. A horrible fear took possession of her that it was too slowly, and that they both would reach Dave's train before she stopped the runaway. She changed the speed, and let the engine gain on her faster.

"I can signal for the siding, if I fail," thought Brier Rose. "Joe will obey my signal." But she shuddered.

Mile after mile they traversed in the direction of Weeping Willow and Red Valley.

In sight of Weeping Willow at last! The Forty-four whistled frantically. Rose signalled for a clear track, and only a train-length apart the Forty-four and the runaway flew past the little station platform, crowded with every man, woman, and child in town, who cheered the flying engine and the white-faced girl like mad.

Joe understood her plan now. He bounded into the station, frenzied with excitement, and telegraphed to Red Valley what Brier Rose was doing; then, from sheer nervousness, he squeezed Foxy until he yelped wildly.

Out of sight of Weeping Willow, and Dave's train in the distance! Nearer and nearer came the runaway. The Forty-four snorted in defiance of being caught. Rose braced herself for the shock. Crash! came the pilot of the runaway into the unprotected rear of the gallant Forty-four. They separated with the shock, and bounded together again; but this time Rose had loosened her hold, and the concussion flung her to the floor, with her soft cheek against the cab seat.

Faint from her fall, she gathered herself together and shut off the steam. Then, with the nose of the runaway viciously pushing the Forty-four, Brier Rose crept like a cat over the tender, down over the trembling engine, and on her hands and knees she crawled over to the runaway, up along the boiler side, into the cab, and crashed the throttle shut when the Forty-four was within a car's-length of Dave's train. She had only strength enough to get her hand on the whistle, hear its shrill call, when, woman-like, she fainted.

When she came to herself she was in the Red Valley station. Dave was bending over her, and calling her name with trem-

THE HEART OF BRIER ROSE

bling lips. She opened her eyes and smiled into his face.

"Oh, Brier Rose, how could you do it?" he whispered, with a shudder.

"I did it for you, David—for you!"

LIZZIE LEE'S SEPARATION

LIZZIE LEE'S SEPARATION

"It all comes of so much marryin'," said Aunt Dony Tuggle, pulling her spectacles down from the crown of her head, where they had been winking at the fire, to set the heel in the gray sock she was knitting. "Some folks just seem possessed to marry."

"And *then* they ain't satisfied — not half of them," supplemented Aunt Mary Battle, looking around pugnaciously, as if daring anybody to prove the contrary.

At these discouraging statements, Cousin Mary Lou, the only one in the assemblage who contemplated a further outrage in that direction, shook her head at the strange Cousin Sara, and laid a slim forefinger on her lip.

"Still, in this case, I shouldn't say that marryin' was to blame," ventured Aunt Emmeline Tally, timidly. "Plenty of folks marry and live peaceable."

"*I* should say that the trouble was with Lizzie Lee herself," burst out Aunt Mary Battle, wrathfully. "If she couldn't marry into the best family in the State of Mississippi, and be too proud of doin' it to bring trouble, she'd better have stayed out of it. That's what *I* say. But she always was a headstrong, con-*tra*-ry little minx."

The six rocking-chairs came to a standstill as if by common consent. The shocked silence made itself felt, even upon Aunt Mary.

"I don't think," said Cousin Sophie Moore, gently, but with increasing color, "that I ever heard any one in *our* family called such a name befo'."

"I believe in callin' a spade a spade," snapped Aunt Mary, "and not 'a sweet little shovel,' specially when it *is* a spade, and not entirely free from garden mould, either!"

She was surprised and hurt to find the family dissenting from her. "Besides," she added, significantly, "Lizzie Lee was a Murchison, and we all know what the Murchisons are."

"She *married* a Mo'!" said Cousin Sophie.

That settled it. Aunt Mary said no more, but she shook her head several times, and ordered black Anna Potts to go tell her mother to tell Lucius to tell Amos to bring a back-log, in a voice which made Anna skip.

"The family never had a breath of scandal breathed on it befo'," said Aunt Emmeline Tally. "It has somehow seemed to me, and I reckon this is a judgment on me *for* it, but it has appeared that our family was kind of sacred—blessed of the Lord, I mean, Sist' Mary. The childern have all been strong and healthy, have been tol'ble good growin' up, and have made sensible marriages with families as good as ours."

"Nearly as good," assented Cousin Sophie.

Aunt Dony Tuggle took up the thread of the narrative.

"Then when Cuthbert Mo' married Lizzie Lee Murchison, we all said that while the Murchisons befo' the war never amount-

LIZZIE LEE'S SEPARATION

ed to much, still there was a Judge Murchison 'way back on her father's side, and it wouldn't have made any difference *what* we said, because Cuthbert *would* have her, and nobody would do him *but* Lizzie Lee. Our sayin' she was powerful con-*tra*-ry didn't bother him any. He must have her, and he got her."

Aunt Dony, being large and comfortable herself, laughed largely and comfortably.

"He took her over the heads of half the county, too," added Aunt Emmeline, with dainty, faded pride.

The younger cousins, not daring to interrupt this serious conversation, Mary Lou leaned over and asked Sara in pantomime if she should hemstitch that ruffle; Sara nodded.

"We said all we could to keep him from it," went on Aunt Dony, amiably, rehearsing the story to get as much good out of it as possible, "and tried to make him see that con-*tra*-ry women don't make good wives. Then we took to her, havin' married a Mo', and visited her and treated her just like we do our own kin. And wasn't she the sweetest, prettiest little bride, with those cheeks of hers as red as a rose, and eyes as big as that!" Aunt Dony put her thumb and forefinger together, making a circle something the size of a silver dollar.

"But now," burst forth Aunt Mary Battle again, "here she ups and says that she can't live with Cuthbert, and she won't live with him, and she's left him. That's what she's done. She's left him. And he a Mo'!"

"Mist' Tally says Cuthbert feels powerful bad," said Aunt Emmeline. "He won't eat nor sleep, but just grieves after Lizzie Lee all the time, like she was dead."

"Oh, Sist' Emmeline!" remonstrated Aunt Dony, who kept unwholesome thoughts well away from herself.

"That's what Mist' Tally said," reiterated Aunt Emmeline.

"Oh, but that's a dreadful thing to *say!*"

"Mist' Tally said so," insisted Aunt Emmeline, with the gentle stubbornness of a negative nature.

"If I could just see Lizzie Lee," said Cousin Sophie, mournfully.

"I hope he won't come here!" cried Aunt

Mary. "I couldn't treat her decently. You Anna, did you tell Amos to bring that backlog?"

"Yas'm. An' he done say he'd breng it jes' soon as he hep feed de calves."

"It's too bad of Lizzie Lee." Aunt Dony rocked and knitted and went on placidly with what she had to say. "She ought to have had mo' patience with him. Cuthbert admits that he was wearin'. But, laws! sister, most men are!"

All five of the rocking-chairs assented to this.

"You didn't hear any reason given for her leavin' him, Sist' Mary?"

"No, it came on me like a clap of thunder, right in church. Miz. Haney wrote it on her hymn-book, just above 'I've a message from the Lord, Hallelujah!' And it gave me such a turn that when we stood up I came mighty near singin', 'Did you know that Lizzie Lee had left her husband?' Those words just danced befo' my eyes."

"If Miz. Haney hadn't told it to anybody but you, I reckon folks needn't have known about it yet awhile," sighed Aunt Emmeline.

LIZZIE LEE'S SEPARATION

Aunt Mary looked uneasy, for she was the greatest talker in the county. The others glanced at her dubiously.

"It was Lizzie Lee's turn to sit up with little Mattie Haney Sadday night, and nobody would have thought strange of her stayin' to let Miz. Haney come to preachin'."

No one said anything in reply. Out on the side-porch they could hear the jerky revolutions of the barrel-churn, as black Alice Potts endeavored to earn her right to wait on the table for company by bringing butter under two hours. Her voice, clear and sweet, rose high over the dull pounding of the churn, as she sang:

> "'Rained forty days and forty nights,
> An' washed dem sinners out of sight.
> See dem sinners swimmin' aroun'
> An' cryin' to Noah befo' dey drown.
> Good Lawd done been here,
> An' bless my soul an' gone!'"

They all listened absently, and the last of the fire-logs fell apart in a shower of sparks, leaving only a glowing bed of coals.

"Well, I'll never believe that it was Cuthbert's fault," sighed Aunt Emmeline.

A little noise on the side-porch made Cousin Sophie say:

"I reckon that's Amos now." But when the door opened it opened to admit, not Amos, but the subject of all this discussion, Lizzie Lee Moore.

They all looked at her with eager curiosity, the little, round, soft beauty, who had set the county by the ears with her great brown eyes and the delicate pink of her peachy cheeks.

She looked around at the gathered relatives shyly, knowing that they had come to talk her over and to condemn her—all *his* relatives, not one of hers. Through the open door behind her the blue sky above the horizon made a light background, against which her slim figure looked singularly childish and helpless.

She stood with one hand still on the door-knob, not as if uncertain whether to enter, but as if seeking to know from their faces which of them had condemned her unheard. And in the embarrassed silence they all felt somewhat guilty.

If she had been less pretty, it would have

LIZZIE LEE'S SEPARATION

seemed brazen in her to come. As it was, her childishness made it appear brave.

Cousin Sophie was the first to recover herself, and to remember that a guest stood on her door-step, waiting to be welcomed. From the time she took Lizzie Lee's hand in hers and felt it tremble, the best in the house was at the disposal, for as long as she chose to remain, of the girl who had left her husband—one of their family, too—and she a Murchison.

"You all are mighty kind," said Lizzie Lee, from the depths of Aunt Mary Battle's rocking-chair, where she had been forcibly placed by that remorseful lady, while she went to take it out on Amos about the back-log.

The December sun streamed in warmly, and the holly bushes tapped their prickly leaves against the window-panes in sharp remonstrance, as if to say, "There are two sides to this question."

"Mighty kind," reiterated Cuthbert Moore's wife. "I—I didn't expect it—hardly."

Aunt Mary appeared at the door, pushing Amos before her.

LIZZIE LEE'S SEPARATION

"Just look at that fire! Ain't you ashamed of yourself to let the fire go plumb out with Miss Sara here visiting your Miss Sophie, and our sweet Miss Lizzie Lee come in from a long ride, and cold as ice from head to foot. Then here you come with a little bitty back-log that would do for kindlin', tryin' to warm up these pretty young ladies with chips. Go on out of here and bring a back-log now that will bend you double. You hear?"

And as the grinning, shuffling boy shambled out, Aunt Mary sat down, having relieved her conscience and regulated Cousin Sophie's household at one fell swoop.

The big plantation-bell, which hung, poised like a great black bird of prey, from the top of a dead tree back of the house, began to clang forth its summons for the field-hands, and Cousin Lisle Moore, Sophie's husband, drawn thither by news from the swarming coloreds about the place that "Miss Lizzie Lee done come to dinner," came in, and greeted his guest with evident concern. It was plain to be seen that he did not know which way the tide of condemnation flowed, and that he was not going to commit himself.

The silence was beginning to be a little awkward, when he suddenly burst out with a subject which he felt to be safe.

"Cousin Sara, I'm sorry to say that old Isrul can't preach Sunday. Cousin Sara said she was dyin' to hear a colored preacher, Miss Lizzie Lee, and I wanted her to hear Isrul. You know old Isrul Potts, the grandfather of all these black apes around hyer." The three children crouching around the fire, seeing themselves glanced at, ducked their heads and dived behind chairs, whence they were all rapped out again by thimbles on their woolly heads. "You know, wife, what a great interpreter of the Scriptures Isrul is. He studies about it *all* the time, and he does get some of the most owdacious meanings from old texts you ever *did* hyer. But they won't let him preach any mo'. I'm surprised to hyer it, but they won't. I'm sorry about that, too, for I did want Cousin Sara to hyer him."

"Why not, Cousin Lisle?"

He was looking into the fire and smiling

LIZZIE LEE'S SEPARATION

to himself, evidently having forgotten about everything except old Israel.

"Well, he and his wife, that Sallie, quar'l all the time, and now they've separated. Say they never will live together again, and the scandal—" Cousin Sophie brought her foot down on his just here, and then he realized. "What's the manner with your fire, wife? Why don't you keep these lazy, loafing blacks at work? You Lelia! You Anna! Get out of hyer and bring some wood. Amos, you were sent for that back-log three days ago. I'll have you shot at sun-down. You hyer me? You've got about fo' hours to live. Get off that cat's tail, suh! Can't you find any other place to put those big feet of yours except plumb on the cat's tail? Hyer, wife! Hold Tabby till this black piece of awkwardness gets the fire made. Not there, Amos! Why don't you put it on the piano and be done with it? Gimme that log and get out. Start running now, and don't stop till you get to Tuggle's branch. Fall into it, if you want to. Gimme those little sticks, Anna. Now put yours on, Lelia. Shut that door, Lucius, and take

LIZZIE LEE'S SEPARATION

Amos's hat out to him. Anna, you go rest Alice at the churn. Lelia, go tell your mother to hurry up dinner. Miss Sara's about starved to death. Wife, I must feed those calves."

Mary Lou and Sara nudged each other delightedly, for, after scattering the black children like autumn leaves in a high wind, Cousin Lisle bolted, with his face as red as fire. He was a shy man, and anything embarrassing always sent him, no matter what the hour, to "feed the calves."

"He never goes near them," whispered Mary Lou.

Lizzie Lee had glanced around at her in a startled way at his unfortunate remark. The rocking-chairs swayed nervously back and forth, no two in unison. The effect upon the observer was that of a choppy sea.

After he had gone they all began to talk rather incoherently at once. But Lizzie Lee interrupted them. She leaned her trim figure forward from the cavernous depths of the great chair, and said:

"I reckon you all feel mighty bad to hear that Mist' Mo' and I have had a fallin' out."

LIZZIE LEE'S SEPARATION

Her voice trembled a little, and it made them entirely forget that it must have been Lizzie Lee's fault, for nobody ever had quarrelled with Cuthbert before—the great, smiling, easy-going young giant, who worshipped the ground his wife's little feet trod upon, and whose only fault lay in "spoilin' her to death with his foolishness over her."

"I want to tell you all about it. That's what I came for. Not that I ever mean to go back to him. It's gone too far for that. Words have been spoken that I can't forget. But just so that you won't blame me." Aunt Mary Battle coughed ominously. "I can't bear to be blamed," said Lizzie Lee, piteously. Aunt Mary patted her hand reassuringly.

"I don't know exactly how it began this time, but ever since we have been married Mist' Mo' has been queer about one thing. Whenever he has wanted me to do a thing, he always begins by beggin' me to do just the opposite, as if I was the con-*tra*-riest thing that ever lived. And when I found that out, it nachally riled me up just same as it would anybody, Cousin Sophie. You

LIZZIE LEE'S SEPARATION

know if Cousin Lisle wanted you to stay at home from preachin', you'd hate to have him get you to do it by beggin' you to go, now wouldn't you?"

"Yes, honey, I would."

"Well, whenever my will goes one way and his goes another, which I must admit is pretty often, he gets up and says, 'Well, I was warned of this, and I wouldn't believe it, but I see it's true.' He won't tell me what he was warned about, or who warned him, and I can't find out. I can't ask other people, can I?"

Six rocking-chairs began to rock very rapidly, and Aunt Dony ravelled out half her knitting without knowing it. Mary Lou laid down her hemstitching and looked attentively at little Lizzie Lee, as if to learn some of the mysteries of this wonderful married state, which Sara alone was aware that she contemplated entering.

"I can't bear to ask questions of people who don't love Mist' Mo', because it looks like I didn't trust him, and it reflects on his dignity. That's why I came to you all, because you are his kin, and you love him

LIZZIE LEE'S SEPARATION

—you couldn't help it—and if I went home Sist' Katy and all would tell me that if I had taken Nelson Ames, or Totten King, or some of the other boys who waited on me, this wouldn't have happened."

The entire relationship stopped breathing at this; but Lizzie Lee continued:

"But Saturday morning it all came to an end. I hope I didn't do wrong. I wouldn't hurt him for the whole world. I—I've always cared too much for him."

The relatives began to exchange uneasy glances. They had been blaming her for not appreciating one of their family, but this scarcely looked like it.

"I hardly know how it began this time. I've tried ever so hard, but these two days seem like a week to me. It was Saturday morning when we began to talk. I believe it was about religion. You know I'm a Babtist and all the Mo's are Methodists; and while I freely admit to you all that I don't keep my temper very well, Mist' Mo' is aggravatin'; and when I get real mad with him, he just laughs at me and tries to get me good-humored again by teasin' me.

LIZZIE LEE'S SEPARATION

He is a powerful tease, you know, Cousin Sophie."

The brown eyes, which had melted but a moment ago, began to flash forth sparkles, and the pink deepened in her cheeks, making her more distractingly pretty than ever. Sara did not wonder that Cuthbert teased her if it made her look like that.

"When I see how easy some people get along with their husbands, and how patient wives are, I do get ashamed of the way Mist' Mo' and I fuss; but somehow, even when I make up my mind not to care, he says something just *too* much, mostly about babtism, and then I flare up. And Saturday it began the same way, and he said this thing about being warned, and I fired up and said I'd not listen to that again, and that whoever warned him against me were wicked people, and if I ever found out who they were I'd tell them so, right to their faces, and look them good in the eyes when I did it." And Lizzie Lee gave them the benefit of that look, but entirely unconscious that she was thus carrying out her threat.

"I said I was goin'—I meant to say I

was goin' to find out, but he must have thought I meant I was goin' somewhere right then, or he never would have said it to me—I'm his wife—but he sort of laughed and said, 'Well, go. I'll not hinder you!' *Oh*, I was so hurt! I said, 'Do you mean it?' And he said, 'Yes.' 'If I go,' I said, and, Cousin Sophie, my voice choked up so I could hardly speak, and he never seemed to notice—don't you think men are queer that way?—'if I go, I'll never come back.' 'No,' he said, 'never come back. Never come back.' He said that to me. Just think of it!"

The quick tears sprang to her lovely eyes in a way which would have made Cuthbert forgive her for anything if he could have seen it. Aunt Emmeline sniffed a little in sympathy. Cousin Sophie sighed, and Mary Lou absently wiped a stray tear on her ruffle, and began to pull threads in her handkerchief.

"I said to him, 'Do you mean that?' And he said, yes, he did. Then he went out, and I went up-stairs and put on my things. I left everything handy for him, and, it being Saturday, I laid out his Sunday suit, and put the buttons in his cuffs, and laid his blue tie and his handkerchief right by his collar, so he wouldn't miss me the first morning. Then I came away. He told me to. He said never to come back. And I never will. Never, never on this green earth!"

Lizzie Lee leaned back in her chair and little lines appeared around her mouth, as if, when she said a thing, she meant it. Aunt Mary Battle wiped her glasses and looked at her doubtfully, as much as to say, "Were we altogether wrong?"

"Have you seen him since?" asked Cousin Sophie.

"He came to Miz. Haney's Sunday—was that only yesterday?—but I didn't see him. I told Miz. Haney to tell you, Aunt Mary. I knew you would tell the rest."

Just then black Edith put her head in at the door.

"Please, 'm, Miss Sophie, here's Sallie Potts, lake to speak to you just one minute."

"Tell her to come in, Edith. Come in, Sallie," cried Sara, eagerly.

LIZZIE LEE'S SEPARATION

It was inspiration.

"Howdy, Sallie?" they said.

"Howdy, Miss Sophie? Howdy, Miz. Tally? Howdy, Miz. Tuggle? Howdy, Miss Sara? Howdy, Miss Mary Lou? Howdy, Miss Lizzie Lee?"

"We hear bad news of you, Sallie," said Sara, lugubriously.

Mary Lou pinched her arm, but Sara shook her off. Sallie was a powerfully built colored woman, strong and tall, and was old Israel Potts's fifth wife.

"Yas'm, I done been cruelly mistreated," she said, standing before them and crossing her large arms. "I has, foh a fack. I heard dat mis'ble, no 'count nigger done been up hyah, fillin' Mist' Mo' full o' *his* side ob de trouble, an' I jis ups an' says, 'I gwine tell Miss Sophie *my* side.' Trouble between husbun an' wife, Miss Sophie, is dey *own* business, an' nobody got a right to say whedder or no. Dat's what *I* says, an' I knows, I does. I ain't been mah'd as many times as Isrul, but I'se had enough trouble wid de one husbun I hab had to make up foh it. I has, foh a fack.

LIZZIE LEE'S SEPARATION

"Y' all know what a little bitty black man dat Isrul is—little, meachin', sneakin'—well 'm, I won't say no mo' on *dat* head. But he is all ob dat. Yas'm, he sho is. He come a-walkin' in Chewsday was a week ago an' says, 'De Sabbath was made foh *man*.' An' I didn't say nothin', case I see he lef' out de ladies a puppose to rile me up, an' jis to spite him I d'termine not to *git* riled. Den he say, 'Paul say, "Let de ladies keep silence in de chutches, foh it is a shame foh one ob dem to speak in chutch."' An' de Sunday befo' I had testified right powerful in love-feast, Miss Sophie, an' I knowed he was pintedly meanin' *me*. 'Ya-as,' I says, 'an' Mist' Mo' say Paul done say his own seff dat he wasn' inspired, an *I* say dat Paul don' need to go to de trouble ob sayin' so, caze dat one ting *prove* dat he wasn't.' An' den I *lafe* at him good! An' he say, 'Mist' Mo' never said dat.' An' I say, 'You a lie.' An' he say, 'Don' you say a preacher ob de gospil is a lie.' An' I say, 'You a lie.' An' he say, 'Don' you say dat agin.' An', Miss Sophie, I said it agin. Den he thow his Bible at me, an' it miss me, an' I thow de rollin'-pin at him, an' it hit him, an' bung his eye good. I did dat foh a fack. Yas'm, I sho did. An' he went an' tole everybody in de chutch dat I half kill him, an' dey come an' try to make us up! An' it done made me so mad, Miss Sophie, dat soon as dey gone I tuk Isrul an' I tied him wid de clozeline to a kitchen cheer, hand an' foots, an' I tuk my sunbonnet offen ob de nail, an' I says, ' I leaves you, Revrun Isrul Potts, an' I'se never comin' back. Fom dis time, you yearns yo' cawn-meal an' bakes yo' bread yo'seff. I'se done wid you.' An' he says, 'Goody!' An' I slams de do', an' he hollers, 'Goody,' an' ' Don't you never come back,' tell I git clar out ob hyahin'."

"How did he get untied?" asked Mary Lou, breathlessly.

"Well 'm, he wiggle hisseff to de do', an' got dat open wid his teef, an' den he wiggle hisseff, still in dat cheer, out to de big road, an' set dar, hollerin' fit to bust, tell somebody come by an' ontied him.

"Now, Miss Sophie, *I* says dat any lady en my place would 'a' done jis same as I

LIZZIE LEE'S SEPARATION

done. Miss Sophie, he had no bizness to say, 'Don' you say dat agin,' when he *know* I gwine say it! Well 'm, I done said my say, an' put my case befo' de white ladies, lake I done said I would. But, Lawd! I don' know what Isrul gwine do widout me. He can't yearn nothin'. But no seff-respeckful cullud lady gwine take in washin' to feed a man what ain't satisfied to feed huh wid bread f'om Heaven by little bits, but what thows the whole Bible at huh to oncet. Dat's all. Miss Lizzie Lee, you lookin' mighty white an pretty to-day. Please, 'm, Miss Sophie, kin I hab a little coffee an' a teeny little bit ob cawn-meal tell I gits de nex' week's wash-money? Thank you, ma'm. Good-evenin', Miz. Battle. Good-evenin', Miz. Tuggle. Good-evenin', Miz. Tally. Miss Mary Lou. Miss Sara."

"And Israel says," Sara began, gayly, "that he has done with Sallie for good; and when Cousin Lisle asked where he got his authority, he said the Scriptures say if thy right hand offend thee, cut it off, and a wife is a man's right hand, so he has cut her off."

LIZZIE LEE'S SEPARATION

Lizzie Lee got up and walked to the window.

"How dreadful such quarrels are!" sighed Cousin Sophie.

"They are perfectly absurd!" cried Aunt Mary Battle.

Lizzie Lee turned around and rested her hands behind her on the window-sill. Her face was scarlet, and her brown eyes drooped. Had she seen and understood? She looked around the room for some way of escape, but they were all between her and the door.

"Cousin Sophie, I'm so mortified. I never thought— It was such a little thing we quarrelled about. He's been so good to me." She raised her eyes from the floor and went on hurriedly, with strong excitement growing in her voice. "It was such a foolish thing— I'll tell him I am sorry. I can't stand it without him. There he comes now, with Cousin Lisle. Cousin Sophie, I'm goin' to meet him!"

And she sprang past the rocking-chairs to the door.

MARY LOU'S MARRYIN'

MARY LOU'S MARRYIN'

With a final exhausted puff and a steamy sigh the engine stood panting in front of the Holly Springs station, while the passengers trooped from the train for their breakfast.

The conductor, carrying a lady's travelling belongings and followed by a girl in a black fur-trimmed gown, craned his neck over the crowd of men in search of one to relieve him. The girl gazed about with a small wrinkle of perplexity in her smooth, white forehead, saying, vaguely,

"I am expecting my cousin to meet me, but he does not seem to be here."

At these words, a man came forward with such a cordial, welcoming smile that her anxious face relaxed. The conductor transferred her affairs to this man, who said:

"It's a leetle too early to expect your cousin yet, but he'll be here. And, in the mean time, just let me take you to the parlor, and I'll make you right comfortable. He'll be here directly. Take this rocking-chair. I'll pull it up close to the fire."

She took off her coat to be able to stand the fire at all, enjoying the cordiality of his manner and the friendly way he beamed on her. She thought it exceedingly kind in Cousin Lisle to have sent word, and, although travelling for the first time alone, she did not feel uneasy at being left solitary in a railroad hotel parlor, for everybody looked pleasantly at her, the women who passed nodded and smiled, the men admired without rudeness, and the warm midwinter sunlight poured radiantly in at the broad windows, as a finishing touch of cordial greeting. It fell upon her white face, revealing that its pallor was not natural, for there were dancing lights in her warm, brown eyes, and mirthful curves in the lips which would not stay pressed together as firmly as she strove to make them. It was a sensitive face, with a certain challenge to Fate in the mutinous eyes and spirited nose, but the fine lines of suffering in her forehead and around her mouth showed that Fate had for the time being triumphed. It showed, too, in the tired way in which she drooped when she was alone. The proud lift of her head was gone, and her eyes saddened with the terrible pathos which looks from the windows of a woman's soul when there is no one but God to see.

She leaned back and absorbed the soothing calm of her surroundings gratefully. The soft Southern voices had no obnoxious Western *r's* boring their way into innocent words, and clinching them with a last twist as final as a nut on the end of a screw, with which to rasp her sensitive ear. The admiring approval which every Southern man bestows upon a handsome girl held no trace of impertinence. It was a different social atmosphere which floated in upon her from all around, and if it brought back stinging memories and hungry yearnings for something inexpressibly dear, now gone out of her life forever, no one should ever know.

"May I ask the name of the cousin you

are expecting?" asked her new friend, appearing at the door.

"Why, I thought—" she began, aghast. Fearful stories of what she had heard of confidence-men flashed into her mind. The train had long since glided down the glistening tracks towards New Orleans. In a moment she was as thoroughly nervous as only a woman can be.

"Mr. Moore," she said, timidly

"Which Mr. Mo'? Holly Springs is full of Mo's."

"Mr. Lisle Moore, of 'The Hollies.' He is my father's cousin, and if you will only put me in the way of getting to him I shall be greatly obliged to you. I thought, when you said it was too early for him, that you knew him, and that he must have sent word."

"Mr. *Lisle* Mo'. Oh-h-h yes, I see. I do know him. It's just as I thought. I knew you must be expectin' to meet somebody from the country, for there wouldn't be no reason why town-folks wouldn't be here to meet you. No, he 'ain't sent word. But he'll be here. No, indeed; you can't have a carriage and start out ten miles to

MARY LOU'S MARRYIN'

his plantation, with only a negro man to drive you. I reckon Mr. Lisle Mo' would whip me if I should let you. You just stay here and go on readin', and I'll send Avery Tuggle."

He disappeared, and presently another man slowly propelled himself into the room, holding his hat in both hands, and smiling protectingly down upon her. Presumably this was Avery Tuggle; but who was Avery Tuggle, and why should he have been sent to her?

"Mr. Mo' will send in for you, that's certain," he said, evidently possessing the points of the case. "It may be about noon before he comes. Roads kind of bad now, but he'll come."

"But can't I hire a carriage? Haven't you a livery-stable in Holly Springs?"

He laughed pleasantly.

"Yes, ma'am, I keep it myself. Oh, I could get you a buggy. 'Tain't that we haven't got none. It's because I wouldn't like to have you go that way, you bein' kin to Mr. Mo'. No, ma'am, you can't start alone. Mr. Mo' would wear me out for lettin' you. But if he don't come by noon, I'll carry you myself."

Sara allowed herself to be adopted by the livery man and the hotel-keeper without further protest. Avery Tuggle came back to say that he would telephone Henry Battle if he saw any of the Moores pass his place. Sara supposed that Henry Battle must be her first owner.

An hour more of solitude drove her to think hungrily of Cousin Lisle Moore's breakfast-table, and the pretty little Cousin Mary Lou, whose wedding she had come to attend. The next time Henry Battle appeared, she ventured to suggest breakfast.

"Of course. It may be right late before anybody comes for you, and you'll be powerful hungry. Just come with me. Leave all your things. They will be perfectly safe. Not much use for furs in this country."

The men lounging around the clerk's desk glanced at her in pleasant approval as she passed. No one in the great empty dining-room except herself. Dozens of negro waiters going to and fro, and behind the swinging-doors such a chorus of sweet voices singing, "Steal away! Steal away! Steal away to Jesus!"—negro voices, so plaintive and sweet that they brought tears to her eyes. A high, clear tenor led the way; great soft, rumbling bass voices cradled the others in a thick blanket of warm melody which swung and rocked Sara's sensitive, susceptible, music-loving soul, and flushed her cheeks with the rapture of it. She awakened to a realization that life still held something for her when such music was occasionally to be heard, and when such a breakfast was being spread. She smiled and then deprecated and finally demurred, greatly to the delight of the friendly waiters, who put a breakfast for five hungry men around her modest plate. And still the doors swung open, and still the smoking dishes appeared. Quail on toast; fried chicken, crisp and golden, with crumpled brown edges and luscious juices of being done to a turn. Waffles as delicate as eggshells; honey as clear as topaz; coffee like amber, with a fat jug of yellow clotted cream; beaten biscuits, daintily browned, to say absolutely nothing of the fruit she was supposed to begin with.

Sarah never forgot that first breakfast in the South, partly because she was enough of an honest woman to declare that she loved good things to eat, partly because of the adorable singing which went on in the kitchen, only interrupted by orders from the head waiter or an uproarious, infectious burst of negro laughter, even more enchanting than the singing, and partly because a vision of loveliness in the shape of Mary Lou rushed in upon her just as she was finishing and gave her the warmest and tenderest of welcomes.

Mr. Henry Battle smiled and rubbed his hands with pleasure. All the men from the office gathered around the door, the negroes crowded in from the kitchen, smiling and nodding with sympathetic delight. Their respectful interest in Sara's affairs was not to be resented, but cordially appreciated. Mr. Battle put her belongings into the carriage, asked Mary Lou about all of the family, congratulated the well ones, recommended new medicines for the sick, obtained her news of the bad roads, felicitated the country in general upon the rise in cotton, and put both girls into the carriage with brotherly solicitude. Sara felt some trepidation in offering to settle a bill with such an intimate friend of hers as he had now become, but he accepted the amount with the same gallant air with which he had attended to her wants, probably regarding it in the nature of a loan.

Mary Lou, with affectionate severity, urged upon her negro driver the necessity of getting home before Christmas, at which he ducked his head and grinned delightedly. A little black girl, who would not ride on the front seat with 'Rastus, crouched at their feet, gazing into Sara's face with the unblinking stare of a young owl.

In this fashion they drove up to the post-office, where Mary Lou went through similar questionings from the postmaster. He was introduced to "Cousin Sara," and adopted her into his family as joyfully as had Henry Battle and Avery Tuggle. Every known specimen of December magazine was in the armful of mail which Mary Lou stowed under the seat, and there were letters—letters of a thickness that warmed one's heart to see.

They drove skilfully between cotton bales piled high, and load upon load of cotton seed gathering around the court-house, and at the edge of town they drew up before the livery-stable. Avery Tuggle, an old friend by this time, hurried out and shook hands cordially.

"Papa will send in for Cousin Sara's trunk, but you keep her check. Give him your check, Cousin Sara. He said he would send for it some time this week. He's ginnin' all his cotton to take advantage of the rise. Good-mornin'."

"Good-mornin', Miss Mary Lou," responded Mr. Tuggle. "Mind you take good care of Miss Sara!"

"Are we all related, Mary Lou?" asked Sara, mischievously, as they drove away. She told her experience.

"Laws, no!" laughed Mary Lou. "Papa never would have forgiven Mist Tuggle for lettin' you start alone. There's nothin' funny in that. You see, we never got your letter till this mornin'. We sent for the mail yesterday, but Israel never brought it up to the house till about three hours ago. Papa just *gave* it to him. Said he'd have him shot if he played him such a trick again. Then he told him you were comin' on the 'Cannon-ball,' and would be waitin' there all this time, and made poor old Israel feel mighty bad."

"Every one was so cordial and smiling that I wondered if they knew what I came down here for."

Mary Lou clutched her arm.

"You didn't tell them, I hope?"

"*Tell* them! Don't they know it?"

"Whoo-ee! No, they don't. Nobody does. I wouldn't have them know for anything in this world."

"How can you keep from talking of it before the servants?"

Sara paused in dismay as she met the fixed gaze of the child crouching at their feet.

"Oh, you needn't mind Alice Potts. Papa calls her 'Non Compos Mentis' half the time. She answers to that name as quick as she does to 'Alice.' Just say something about Christmas, and she'll never know. And 'Rastus is deaf. We don't talk these things over very much in the South. If we can keep it a secret up to the very day, we are mightily pleased to do it."

"But why?"

"Well, on plantations, if the coloreds know there is going to be a marryin', they come in crowds. Whole families shut up their houses and come and stay, and you've got to feed them, and doctor them if they get sick, and give them clothes fit to be seen if you don't want to be disgraced. Then," and her face dropped, "maybe the girl changes her mind at the last minute, and it comes a little hard on the boy if it's known."

"Naturally."

"You've never been south before, Cousin Sara?"

"Never so far as this. I spent last summer on Lookout Mountain."

"Oh, that's way up yonder."

"Mary Lou, tell me about the man you are going to m-a-r-r-y," spelled Sara, in deference to the stare of Non Compos.

MARY LOU'S MARRYIN'

Mary Lou laughed and twitched at her ribbons half impatiently.

"Cousin Sara, would you be very much disappointed if there shouldn't be any marryin' after all?"

"After my coming a thousand miles to see you go through with it, and having brought you a white silk dress in my trunk, and a veil and gloves and shoes to match? Oh no!"

Mary Lou gave a little bounce.

"Oh, Cousin Sara, honest true, did you? Did you, Cousin Sara? Go *along*, Bird! You *are* the slowest horse! 'Rastus, can't you make him keep up with Fanny?"

A pink flush crept into her soft cheeks. She was a little puff-ball of femininity, round and smooth and distracting.

"It was to have been—I mean, it is to be on Christmas Day, because we can have all the preparations made and the necessary cookin' done as if for Christmas, and the negroes will never suspect. What kind of silk is it, Cousin Sara? Oh, lovely! I must just give you another kiss right this minute. I was going to marry in a travellin'-dress. I never thought of a white silk."

"When do you expect to have your clothes made?" asked Sara, apprehensively.

"Oh, they are all bein' made at my dressmaker's in N' Orleans. We'll have to make the white silk ourselves. Old Aunt Sallie sews beautifully. We'll tell her it is your dress."

"Well, tell me about HIM — my new cousin."

"Oh," answered Mary Lou, half impatiently, "he is big and fair, with blue eyes, and light hair that would curl if he gave it a chance. He used to be lively and full of fun, but he's so stern and sad now that I am half afraid of him. He won't tell me what troubles him. He says I am never to be worried by anything, and that he owes it to me to make my life as free from sorrow as he can. I suppose you have all the latest New York styles in your head, haven't you?"

"Yes," answered Sara, absently. She wanted Mary Lou to talk further about her *fiancé*, but this she seemed singularly averse to do.

As they drove into the plantation grounds they were assailed by a chorus of dogs and a swarm of black children who flocked from the house, followed by the various members of Mary Lou's family, all eager to bid her welcome. To Sara's surprise, Non Compos, the silent, the stupid, sprang up in the carriage, waving her little black paws and shrieking,

"We brung her! We done fotch de Princess! Keep away from her, Lucius! Don't tech her dress wid yo' black hands, Anna! You, Lelia, drap my Miss Sara's handbaig! I'se gwine ca'y it my own seff."

Sara's enjoyment was infectious. Her cousins reprimanded the black children, whom they spoiled badly, and made a soft clamor over their kinswoman, as delightful as it was unique.

Already the pallor was leaving Sara's cheeks. She felt the intoxication of this new mood which had descended upon her. Everything was queer and different, but she chose to enjoy it in her own way. She could scarcely keep her eyes from the black children, who were constantly under-foot, but who were allowed to stay "in de gret house," to learn of their mothers, the house-servants, how to wait on the ladies. It was,

"Lelia, get Miss Sara a glass of water;" or,

"Alice Potts, pick up Miss Sara's handkerchief;" or,

"Anna, go tell Lucius to get your pretty Miss Sara some persimmons."

Mary Lou persuaded her father to send back to town that same day—an unheard-of thing—for Sara's trunk. When it came, she enjoined the strictest secrecy concerning the white silk. If it had been stolen, and detectives on their track, the two girls could not have looked more guilty or more conscious.

It was, indeed, as Mary Lou had said. The wedding preparations went forward as if for Christmas, and nothing was said about them except in the secret recesses of Sara's room, where, closeted with Mary Lou's mother, Cousin Sophie, endless discussions were held.

Finally, Sara could stand it no longer.

"Sophie, do tell me about the man himself."

"Why, Cousin Sara, I don't believe the child knows which one she's going to marry. But you are right. It certainly is time she decided. I reckon I'll ask her."

Sara gave up after that. She accepted this, too, as a part and parcel of the new *régime*, and really came to enjoy the uncertainty. She made sure of a wedding of some kind by stipulating that the white silk was for a wedding-dress. It was bought for that purpose; it must be used for that purpose — or, in the pause which ensued, Mary Lou had awful visions of the dainty thing being withdrawn from her enraptured sight altogether and whisked back the thousand miles whence it had come.

The girls, although so totally unlike, grew fonder of each other daily. They strolled in the grounds, followed by the devoted Non Compos, and attended by Lucius, clad in a cast-off street jacket of Mary Lou's, whose fulness flapped around his lean shanks with such an old-world air that

Sara named it "Lucius's surtout," to his infinite pride.

So easily are good manners corrupted, or rather are habits of thrift deposed, that Sara no longer wondered why they did not fence the chickens in, instead of converting the violet beds into brush-heaps to prevent the flowers from being eaten bodily. She, herself, now sent the children flying to drive the hens away, or to pile the brush higher. She took a violent interest in the flower pit where Mary Lou hid her more delicate plants from the mild rigor of the Southern winter. She grew to love the great white house, with its green blinds and deep porch. But the great plantation bell, swinging like a grim, black sentinel from the top of a dead tree, possessed a fascination for her which with difficulty she resisted at all.

"I do believe I shall die, Mary Lou, if I can't ring that bell," she said, in one of their walks.

"That's the only thing on the whole plantation that you can't do," said Mary Lou. "It would bring all the hands in from the cotton-fields, and all the coloreds from far and near. It would make any one ridin' along the big road turn in here to offer help, for it means that somebody is wanted mighty bad, or that something awful has happened at 'The Hollies.' It always scares me to hear it, except at the regular times for it to ring, and then it only gives a peal or two. I do honestly think I would curl up and die if I heard it ring a long time."

Sara looked at it longingly, but passed on.

"Miss Sara, take me back up North wid you when you go," cried Non Compos, who was skipping along at her side.

"Why, Alice, what would you do in a place where there are no grounds for you to play in—just houses and houses?"

"No grounds, Miss Sara?"

"No, none at all."

The child stood kicking at a pile of dead leaves in silence for a moment. She always kicked at something when she was about to think. The movement of her feet seemed to loosen a tendon in her brain.

"No grounds, Miss Sara? Then what do de rain fall awn?"

"Just listen at that," said Mary Lou. "I told you she was non compos mentis. Run along, Alice. You bother Miss Sara. Princess, what would you do if you were engaged to the wrong man, and had come to the conclusion too late that you loved the other?"

"I would break off the engagement, and marry the one I loved," said Sara, with sudden fire in her tones.

"Oh, my Brown Eyes! How I love to see them light up!"

"Don't!" gasped Sara. "Don't call me that. I—I can't bear it."

Mary Lou was a little ball of thistledown in many matters, but in affairs of the heart she was a Minerva and a Solomon rolled into one. She looked at her cousin's white face out of the corners of her eyes.

"Some man has called her 'Brown Eyes,'" she thought, shrewdly.

"The man I love has brown eyes," she said, softly. "That is why I speak of yours so often. I can't bear blue-eyed men."

Sara laid her hand on the bosom of her gown, under which, in a flat gold heart, lay the picture of a blue-eyed man. She said nothing.

"She has got his picture in a locket," thought Mary Lou, who watched her. "And, oh dear, there are tears in her eyes."

"I am going back to the house," she announced. "And the children must come with me. Shall we leave you to finish your walk?"

"Yes, do. I will be back soon," said Sara, eagerly.

"She wants to look at that locket, and think about him alone. She needs a good cry," thought Mary Lou, gathering up the children, and going back with a lovely look of sympathy on her sweet face.

Left alone, Sara drew out the flat gold heart and looked into the face of the man who gazed out at her with brave, fearless eyes. She unfastened the chain, and held the locket in her hand. Then slowly she took out a letter, almost dropping to pieces from many foldings, and read it again. How many times she had read it already!

"Although this letter has neither beginning nor ending, you will know from whom it comes. Dear heart, let me call you so, although I have no right to ask, and, what is worse, I never shall have. I am going away without seeing you again. I am strong enough for that, but God knows what might happen if I should beg another interview, even to say the good-bye which wrings my heart to write.

"I must tell you, in desperation I must tell you that I am bound to another, a sweet girl, whom I thought I loved until I met you. But you, *you* have shown me what it is to love. I feel that I am both weak and false to her to tell you this, but perhaps, after last night, I owe it to you to paint myself in my true colors, and let you despise me if you will. The only comfort I have is that your heart is as yet untouched. I beg that you will pardon the arrogance of the words 'as yet,' but I swear to you, by the Heaven above us, that you should love me if I were free to ask! Yet I love you so much that I am glad the suffering is to be all mine. I would only plead for a little regret from you, which the tender light in your brown eyes tells me I should have, because you know my pain.

"I have tried to be true to her, yet this whole letter is a betrayal. I only ask you to believe that I never dreamed I loved you until last night. If I had, I should have gone before now, and this letter would only have come to you sooner. It seems as if Fate first tempts, and then judges inexorably. I have meant to do right, but see how I have fallen. I claim to be strong, yet I deliberately choose to send you this letter with all its fatal weakness apparent, and all my love in its possession.

"I have not the courage to write a last word to you. Let this letter be broken off as abruptly as the current of our two lives. I shall always love you, my Brown Eyes, always, always!"

When she had finished reading, Sara put her face down in her hands and wept bitter tears upon the face in the gold heart.

She was so intense a personality, and so

self-poised, that few would have suspected her of the weakness of the gold heart. Yet the few might have known, might even feel an understanding compassion for the foolishness which wrote "*Toujours*" across the flat, gold surface. Ah me, and why not? Are the self-poised forever debarred from foolish delights? And what would become of the men if only the silly, clinging women were the ones to write "*Toujours*" across a gold heart, and treasure it against their own?

When Sara appeared in the home circle around the great open fire, there were traces of storm in her face to Mary Lou's observant eyes, and there was a pathetic droop in her tired smile which wrung Mary Lou's faithful heart.

"Miss Sara, will you write my letter to Santa Claw?" asked Non Compos, thrusting paper and pencil into her hand. "Miss Ma'y Lou done wrote Anna's an' Lelia's, but *I* want de Prin-*cess* to write mine."

"Well, tell me what to say," said Sara.

"'Dear Santa Claw, thank you for what you brung me las' Christmas, and please, sir, don't forget me dis time. I'se been a good chile. I done kep' Miss Sophie's chip-box full, an' ain't forgot to draw water and ain't been imperent, an' I'se hunted aigs faithful. Please, sir, bring me a gol' ring an' a silk handkercher. An' bring Miss Sophie her good health back again, wot she done los' since col' weather sot in. An' bring Mist' Mo' some new slippahs, 'case his is all busted out, an' it don' look right for a gemman to wear slippahs dat ain't fitten for nobody but a niggah to wear, an' dey would jus' fit Uncle Israel. An' bring Miss Sara—'"

"Say, Miss Sara," said the child, interrupting herself, "ain't you got no sweetheart, lake Miss Ma'y Lou?"

"No, Alice, I haven't any."

"Miss Ma'y Lou's got three or' fo'. Ain't you got ary one? Ain't you never gwi' mah'y?"

"No, I think not."

"Miss Ma'y Lou, why don' you give Miss Sara one of yours? You got mo'n you kin han'l'; I done heard Mammy say so."

"She can have any of mine, except one,"

said Mary Lou, laughing. "I must keep one for myself, Alice — one with brown eyes."

"What color eyes does you want your husbun to hab, Miss Sara?"

"Blue, please," said Sara.

"Now go awn wid de letter. I'se gwi' ask Santa Claw for jus' de ve'y ting y' all want de mos'."

"'An' bring Miss Sara an' Miss Ma'y Lou two husbun's, an' don' get 'em mixed up, 'case Miss Ma'y Lou 'bleeged to hab one wid brown eyes, and Miss Sara got to hab de blue-eyed gemman. An' bring Miss Sara's husbun a red cravat an' a gol' collar-button, 'case Miss Sara's a fine lady, an' she mus' hab a fine gemman to mah'y.'"

"The very thing we want most," murmured Mary Lou. "Consider our pain, Non Compos, to discover that you have divined our secret with such unerring skill. We shall await Christmas day with eager and expectant joy. I sometimes wonder, Alice, if you are as great a fool as you look."

Alice ducked behind Mary Lou's chair,

whence she was promptly rapped out again, with a thimble brought in smart contact with her woolly head.

It came to be the day before Christmas. The house was in holiday attire. Invitations were out long ago for a Christmas party in honor of Cousin Sara. Holly framed the pictures on the walls, and bunches of mistletoe swung from the hanging lamps, making even the stateliest progress through the great rooms dangerous. In the yard, for days the large brick ovens had been harboring goodly sights and exhaling goodlier odors, the results of which were now stacked in kitchen and storeroom. A huge box from New York arrived during the day for Sara, with which she closeted herself in her room, demanding that no one should enter there on pain of death. She only emerged to take her meals with the family, and had she not been unusually preoccupied she would have noticed Mary Lou's pale cheeks and mournful droop.

In the early twilight she called Mary Lou into her room. Everywhere stood odd-shaped parcels, covered with newspapers to conceal their identity. Sara lighted the candles in their sconces and began to dress.

"Well, I've done it," said Mary Lou.

"Done what?"

"I've broken my engagement."

"For the love of Heaven!" exclaimed Sara. "When?"

"Just now. That was he you heard ridin' out a little while ago. Perhaps I ought not to have let him come."

"When did he come, and why didn't you tell me?"

"He only got in town this mornin', and he's going on this evenin's train—to New York, he said, but what he's going there for I don't see. I am just as mad as I can be, too."

"What are you mad at? You've been wanting to get rid of him ever since I came. Was he broken-hearted? Did he reproach you and storm around?"

"No," exclaimed Mary Lou, explosively. "That's just what is the matter. He only said, 'Are you sure? Oh, Miss Mary Lou, are you sure?' And when I said 'Yes,' as kindly as I could, for fear of hurtin' him, he grabbed my hand and nearly crushed it, and seemed glad—actually glad. His face was perfectly radiant."

Tears of mortification dimmed Mary Lou's eyes.

Sara finished her hair in a hurry, warmed by Mary Lou's narrative, and struggled with the waist of her gown. One of the many hooks thereon caught in the invisible chain which held her locket and snapped it, dropping the gold heart on the floor at Mary Lou's feet, where it lay face upward and open. She glanced down, then seized and looked at it.

"What are you doin' with Mr. Hilary Bonner's picture in your locket?" she said, in a low tone.

Sara almost gasped.

"How did *you* know his name?" she said.

"Why shouldn't I know the name of the man I was going to marry, who rode out of these grounds not half an hour ago?"

Sara's face grew wan even as her cousin looked at her. They stared at each other in a silence too deep for words.

"He has been here," whispered Sara, "and I never saw him? Oh, Mary Lou, and are you the girl I've been envying all this year?"

"And are you the girl he was so glad to be rid of me for? Oh, Cousin Sara, I'm so glad! I'm so glad! He is the best and the bravest and the truest man in the world—except one."

"But he's gone," whispered Sara again, still holding the gold heart in her two hands, and staring at the face with a look in her eyes that Hilary Bonner should have seen.

"I wonder, if I sent after him, if anybody could overtake him. Saladin is lame, and papa and mamma took the carriage-horses right after dinner to go for grandma. It's moonlight, so they won't be back till nine o'clock. Lucius couldn't catch a fly if I sent him. Oh dear! oh dear! and he was going on the evenin' train!"

"Do you suppose he still cares for me?" asked Sara, timidly, searching Mary Lou's excited face wistfully.

"Am I sure? Of course I am. If you could have seen him! And his going straight to New York! Oh, it's like a book —it's like a play! I'll take Fanny and go after him myself."

"Lucius took Fanny and went to town for me," said Sara.

"Then we can't possibly overtake him. We'll just have to sit still and let him go. And he breakin' his heart for you, and you breakin' yours for him!"

Mary Lou wrung her hands.

"The bell—the plantation bell!" cried Sara.

Mary Lou started and hesitated.

"There's no one to be scared by it, and —who cares if there is?" she cried, in glad defiance of a law of her childhood. She sprang up, and the two girls rushed from the house and made straight for the dead tree from which swung the black sentinel of "The Hollies."

Mary Lou seized the bell-rope, but Sara laid a detaining hand on her arm.

"Oh, Mary Lou, suppose he shouldn't want to come back," she said, with strange timidity.

"He needn't know that you want him.

I'll tell him I did it—that he forgot his gloves!" cried Mary Lou, stoutly. She leaned to the bell-rope, and a startled peal cut the crisp air, which sent a tingle along Sara's nerves like electricity. Louder, faster, gayer swung the bell. The negroes swarmed in from their cabins. Lanterns flashed to and fro in the soft gloom, and overhead the pale moon was riding high in the heavens, growing brighter and clearer as the shadows deepened upon the earth she watched.

"Wha's de mattah, chile? Wha' foh you ring dat bell?" they cried, gathering around her.

"Because to-morrow's Christmas!" cried Mary Lou, growing more reckless with the intoxication of the sounds and life she stirred with her clangor.

A burst of wriggling laughter answered her.

"Dat's so! 'Fo' Gawd, de chile's right! Ring 'em out, Miss Ma'y Lou. Hit sho do soun' putty to hear de Christmas bells a-ringin' in de ole plantation groun's once mo'!"

They scattered to their homes again, as satisfied and delighted as children.

"We are ringin' our weddin'-bells, Cousin Sara!" cried Mary Lou, ringing more madly than ever.

"Ours?" exclaimed Sara.

"Yes, I'm going to marry the brown-eyed husband that Alice ordered for me."

"Hush a minute," urged Sara.

Mary Lou paused. Sara was shaking like a leaf. "Will he know that we want him if he hears it?"

"Of course he will. So will everybody else. I only hope that a dozen others won't happen along the road just now."

"Listen!" cried Mary Lou. "I heard a horse — but there are wheels, too. That isn't Hilary. He was on horseback."

They listened eagerly. Soon a buggy, with the horse lashed into a gallop, came into view, turned in at "The Hollies," and Lucius, his face ashen with fright, tumbled out, while poor Fanny heaved and panted as a petted horse always does if she ever goes out of a trot.

"Foh—foh de Lawd's sake, Miss Ma'y

Lou," stammered Lucius, "wha—wha's de mattah? Is Mist' Mo' daid?"

"No, nobody is dead. But to-morrow is Christmas, and I knew you'd come crawlin' home about New Year's, wantin' to hang up your stockin' if I didn't hurry you up some. Go 'long and feed Fanny—mind you rest her and rub her down first. I'm going to ring the bell some more."

"Oh, don't!" cried Sara, almost in tears. "He's too far away. He can't hear."

But Mary Lou's spirits were now at a pitch when she wanted to ring the bell at any cost. She swayed to and fro, bending her weight to bring forth a great, sonorous clangor, which made Sara mute with apprehension.

Presently she stopped of her own accord.

"Now, if he hasn't heard by this time, it is too late," she said.

They stood hand in hand, listening with bated breath.

"I hear him," whispered Mary Lou, pulling at her cousin's hand. "Don't you?"

"No."

"Now do you?"

MARY LOU'S MARRYIN'

"I hear somebody, but perhaps it is not he."

"But I know it is. Hear him! Nobody except Hilary can ride like that, and over these roads. *Hear* him ride!"

Mary Lou was quite wild with excitement. Some of it subtly communicated itself to Sara. She seized the white shawl from Mary Lou's shoulders, and, flinging it over her head, she sped down the drive to the big gate, which Lucius had left open.

She stood trembling between hope and fear. Nearer and nearer came the hoof-beats of a madly ridden horse. She looked down the road, and in the pale moonlight she saw the figure of his gallant rider, a part, a parcel of the swift motion beneath him. The horse, dripping with foam, swerved in at the gate, and at the sight of Sara the rider flung himself out of the saddle, splashed with mud from head to foot, almost at her very feet. He had expected to see Mary Lou, but terror at the bell's summons and all other apprehensions merged into his incredulous delight at being confronted instead with the deep eyes whose rays had fired his heart on Lookout Mountain, when he had first loved her and had written that letter.

They remembered to come into the house at last. Mary Lou met them with a whimsical smile lurking in the corners of her mouth.

"Howdy, *Cousin* Hilary?" she said, demurely.

Hilary Bonner laughed in proud delight at Sara's furious blushes, and caught both of Mary Lou's hands in his.

"She has been telling me about it. It is all your doing, you dear, generous little woman."

The quick tears sprang to Mary Lou's eyes. She turned away to hide them.

"For mercy's sake, Cousin Sara!" she cried. "You've ruined your dress. It's mud from head to foot. How in the world—" She paused, and glanced at Hilary, but he was studying the holly branches on the wall, and Sara had discreetly vanished.

On Christmas morning the air was thick with exclamations of "Christmas gif'!" "Christmas gif'!" as Alice and Anna, Lucius and Lelia built the fires by candlelight, and forced the family to eat breakfast at dawn, in order to hasten the time when they would be "Santa Clawed."

The servants, old and young, with no end of unexpected relatives and camp-followers, gathered in the parlor to view the mysteries of the bundles hidden therein. As per previous announcement, the children's work for the day had all been done, churning finished, wood and water brought, everything cleared away, hearths painted, and a roaring, picturesque wood fire built in every room. The chair which held their stockings was uncovered, and with a howl of delight they sprang upon their property and undid it, "just as if," said Cousin Lisle, "they had not had the identical things every Christmas since they were born."

The family were in gales of laughter over the antics caused by the wonders contained in Sara's box. How appropriate a blue-gauze fan seemed to old Aunt Sallie, and a card-case to the mother of the twins! Every one got what had been asked for in the letters to Santa Claus.

Presently Non Compos wrenched her gaze away from her precious gold ring and pulled at Sara's hand.

"Miss Sara," she whispered, "I don't see no husbuns."

"Go out and look, Alice. Ask Cousin Lisle if he saw any."

She came back radiant.

"He's out dah talkin' to Mist' Hilary Bonnah an' Mist' Clinton Millah. He say he b'leeve he did see 'em skitin' aroun' some befo' breakfas'. He say he reckon dey roost in de trees wid de tuckeys, an' he gwi' hab 'em shuck down foh me, soon's I fin' him de tree dey's in. He say he b'leeve it's jest gwi' to *rain* husbuns at 'De Hollies' dis day, an' Mist' Hilary Bonnah an' Mist' Clinton Millah lake to bust dey sides a-laffin' at him."

Towards evening the gathering of the clans began. Buggies and carriages filed into the grounds, negroes ran about shouting contrary directions, caring for horses, removing wraps, shaking out tumbled dresses, supplying pins, even needles and thread when needed. Mary Lou was not to be seen.

MARY LOU'S MARRYIN'

Everybody met Cousin Sara, who, in a pink gown which matched her cheeks, had a strange fashion of disappearing every once in a while. Mr. Bonner's tall, soldierly figure also had a way of being in unmistakable proximity to the pink gown, and at the sight of his bronzed face and intent gaze the pink flush always deepened.

There were lights and music and flowers. There were fresh, sweet faces, and soft, melodious voices, much laughter, and a general air of Christmas gayety in the holly-trimmed rooms. And suddenly a hush, when out from among the throng, which parted to give them place, walked Mary Lou in her sweet white silk, with Mr. Clinton Miller, who paused in the centre of the drawing-room floor, just under the most generous cluster of mistletoe. And in the surprised silence the preacher's voice was heard in the solemn words which gave Mary Lou another's name, even as she long ago had given her heart into another's keeping.

A brief moment, and Mary Lou's marrying, so admirably kept secret, was *un fait accompli*, a thing to talk about and to emulate for years to come, in its secrecy and dispatch, and complete fulfilment of all requirements.

Everybody, including Mr. Hilary Bonner, looked so radiantly happy that sounds of woe disturbed his kindly heart. He sought out the mourner, and discovered poor little Non Compos on the porch in the dark, sobbing as if her heart would break. All day long she had wandered around unnoticed, gazing into the tree-tops, whence she had the solemn assurance that the promised husbands would come. Mr. Bonner drew forth the story of her deception.

"An' if dey did come now," she wound up, with a final burst of tears, "it's too dark for me to fin' 'em. I didn't see Mist' Millah drap, an' Miss Ma'y Lou done foun' him widout me. Now, if Miss Sara's comes, I cain't see him, an' I did want to take de Prin-*cess* her husbun wid my own hand."

"Poor little soul!" said Mr. Bonner. "Cheer up, Alice. You can take me! There is Miss Sara standing in the door. Let's go to her. I'll be her husband, if she will let me."

MARY LOU'S MARRYIN'

Alice sprang up enchanted, and dragged him into the light which streamed from one of the windows.

"Has you got blue eyes?" she demanded. "De Prin-*cess* p'intedly said her husbun mus' hab blue eyes."

"Bless her dear heart!" murmured Mr. Hilary Bonner fatuously.

Hand in hand, the two oddly assorted companions walked up to where Sara stood alone. Alice's face, polished by her tears into a glister resembling a kitchen stove, was resplendent with generous enthusiasm.

"Dear Miss Prin-*cess*," she whispered, "Santa Claw sont you a husbun wid blue eyes, lake I done axed him!"

THE STRIKE AT THE "BILLY BOWLEGS"

THE STRIKE AT THE "BILLY BOWLEGS"

I

THE miners of the "Billy Bowlegs" had struck. At twelve o'clock, on the 20th of December, two hundred of them, regardless of the women's prayers and tears, walked out. To be sure they might starve before the owners of the mine, Boston men, would consider their claims and settle, for Cactus Camp was only a mining hamlet in Southern New Mexico; but the miners had ill-considered that possibility. The week before, they had been visited by the editor of the *Labor War Cry*, who had made several speeches, in which he set forth their wrongs. They were the sort of men who nurse their wrath against capital for no other reason than that they are poor and others are rich. Unreasoning hatred against a class of whom personally they know absolutely nothing smouldered continually in their minds. Anybody with a tongue in his head could stir it up. It was always ready and willing to leap into a flame at a breath. Perhaps with these men of the "Billy Bowlegs" the sight of the rich ore they mined was a constant reminder that there were people in the outside world who revelled in the gold and silver their toil produced, and of which they received such a meagre share. But, more than all, it was the native anarchy which forever breeds in the souls of a certain class in America, of which the miners of the "Billy Bowlegs" were shining examples.

In this instance they considered that they had a hold upon the mine-owners which would serve to bring them to book. It was that they had just struck the richest vein of ore yet discovered in the rich "Billy Bowlegs." In these hard times, Peters, the ringleader, urged upon the men that the mine-owners would not be so foolish as

STRIKE AT THE "BILLY BOWLEGS"

to stand out on the fifty-cents-a-day raise which they demanded when this new vein had just caused the stock to advance one hundred and ten points.

There were but two men in Cactus Camp whose fortunes would not be affected by this strike. One was the telegraph-operator, and the other was the clerk in the general store which purveyed to all the wants of the miners.

The third day after the strike had been declared, Carson, the clerk, closed up the store and crossed the street to the telegraph-office. He often wished that this telegraph-office, which was the one thing in camp which was not in the grasp of the store, could have been placed there, not so much from the idea of greed, not because his department-store desired to become the one monopoly of Cactus Camp, as because he wanted company. There were long, lonely nights in that isolated mining-camp, and many of them when the friendly clicking of the companionable little machine would have been almost as welcome as a human voice.

STRIKE AT THE "BILLY BOWLEGS"

Carson, although perhaps the loneliest man in camp since his wife and baby died, had been one of the most social. His whole nature craved companionship, the companionship of woman, and, failing that, the companionship of man. He was said to be "down on his luck." No one knew his personal history, except that he and his brother, in prospecting, had discovered the "Billy Bowlegs." The ore on the surface was not particularly compact, but, even scattered though it was, it was so rich that they considered that it was decidedly worth while. They staked their claim, put all their money into it, and took nothing out. All the ore seemed to be on the surface. Disappointed, yet hopeful with the hopeful fever of mining, they borrowed money, and went deeper — deeper into the mine, and deeper into debt. For six years they wasted their time and their money in the mine, until finally, disheartened, they offered the mine and all their mining property to the banker who had lent them five thousand dollars. He demurred. He did not want the mine. He did want his money. He had more faith in it than they. He offered to extend the loan, and let them have more, but they refused. So they sold the "Billy Bowlegs" for five thousand dollars. The very next week—not the next year, nor in ten years, but the very next week—the new owners struck a vein which yielded a hundred dollars a ton, and within a year the stock of the "Billy Bowlegs" was the most valuable in the market, and the share-holders were on the high-road to fortune.

People forget easily. Few, even of the miners who had heard this story, remembered that Tom Carson, who lived so silently in their midst, had just missed fortune by a hair's-breadth. Carson was the only man in camp who had good reason to hate the capitalists, who were gathering in the millions which should have been his, or who bore in his own soul a justification, if such there be, for the anarchy which poisoned the camp. But, curiously enough, he was the one man who was free from the taint. Even Scott, the telegraph-operator, who kept his views to himself, was thought to be friendly to the strikers.

STRIKE AT THE "BILLY BOWLEGS"

On this evening Carson went over to see him. The machine clicked pleasantly on the table.

"Hullo, Scott!"

"Hullo, Carson!"

"Sort of nice, friendly, homey sound the little fellow has," said Carson, jerking his head in the direction of the machine. "Seems the next best thing to hearin' the tea-kettle a-singin' on the stove."

"Hum! Guess you wouldn't say it sounded like no tea-kettle if you heard the message it ticked off just before you came in. It's to Peters, in answer to the letter he sent the boss three days ago."

"What does he say? Does he give 'em the raise?"

"Raise? Not much. He gives 'em the devil. He says, 'Troops from Fort Bliss ordered to Cactus Camp. Absolutely no other reply. Withdraw demand or fight.'"

"Good Lord, you mustn't give 'em that message!" exclaimed Carson, anxiously. "It'll be like a spark to powder. They are just crazy enough, with the whiskey and the women cryin' and beggin' 'em to back down, to fight like Injuns."

"Not give 'em the message? I *got* to give it to 'em. I can't keep the thing back. Besides, what's the use? My not tellin' 'em wouldn't keep the troops away."

"That's so," said Carson, thoughtfully.

"Seems to me that's a mighty short, sharp message to send the boys, slap-dab, without any temporizin' or soft solder first to let 'em down easy. It'll make 'em crazy."

"I suppose it's all along of the Leadville and Cripple Creek troubles. The boss means to begin where the other fellows left off. Well, it will only end things for everybody quicker."

Scott glanced at Carson hastily. Carson's chair was tilted back against the wall, his chin was bent forward on his chest, and his hands were dug down deep into the pockets of his trousers. He was staring moodily at the floor.

"Sorter down on your luck to-night, old man, aren't you?"

"'Tain't my luck," returned Carson, moodily. "I was thinkin about Jim."

STRIKE AT THE "BILLY BOWLEGS"

"What about Jim?"

"Don't you know? Why, two weeks ago he sent for his wife and baby. He's been savin' up for a year to do it, and she at home in Maine, waitin' to get word to come. She packed up the minute she got the letter, sold all the stuff except what she was goin' to bring, and started the very day of the strike. She ought to be here to-morrow. Just think what she's comin' to! Jim out of a job, and so sick he can hardly stand—you know he's got consumption, and maybe she's comin' here to starve."

"Lord, that's tough!" exclaimed Scott. "If there is to be fightin', she stands two chances of bein' a widow. Don't see how she can miss it exactly."

"The baby is just three years old, Jim says," added Carson.

"That so?" said Scott, vaguely.

"Mine was just that old when he died," said Carson, gently.

"Didn't know you ever had one," said Scott, awkwardly.

"Didn't you?" said Carson, raising his head. "Oh yes, you did. You must have known, but you've forgotten. It doesn't mean much to a man when a baby dies, unless it's *his* baby. Why, the mine's named for him! My wife named it. She had a sight of fun in her. I used to call the little chap 'Billy Bowlegs,' because his little legs weren't quite straight when he was born—they grew straight before he walked, though—and Katie said, when we struck ore, 'Let's call it the "Billy Bowlegs." It may bring us luck.' So we did. She was the merriest-hearted little woman you ever saw."

"What's become of *her?*" asked Scott, sympathetically.

"Dead," said Carson, shortly.

Scott was silent from embarrassment. He had never heard Carson talk in this way before.

"Yes, everybody belongin' to me is dead. As fur as I am concerned, it makes no difference whether this strike is on or off. My job is secure, and so is yours. It's just my luck to be safe from harm now that I'm all alone, and nobody to share it or care whether I live or die. There's Jim, now. I declare I'm a little tender when I think of Jim's wife and the baby. I wish we could do something."

"Do you know what night it is?" asked Scott.

Carson glanced at the calendar.

"Christmas eve!" he exclaimed. "Christmas eve! Four years ago we were just gettin' ready for the biggest Christmas tree for our little feller, with candles on it, and cranberries and pop-corn enough to set a child crazy to look at it, an' he never lived to see it. Took sick and died as quick as that. I've always wished I could see some little fellow's eyes shine over a Christmas tree like we were goin' to give Billy. Do you think we could get up one for Jim's baby?"

"I s'pose we could get it up all right, but what would be the good, with him out of a job and the strike on?"

"Well, if you'll just hold that message over till to-morrow night, we could give 'em one Christmas that they'd never forget, even if it was over most before they knew it. Will you?"

"I don't know as there would be any

harm," said Scott, dubiously. "They can't send anywhere for arms or ammunition. We're too fur away from the railroad. I don't see who could be hurt by it."

"I don't, either. And it does seem a pity to spoil Christmas. It comes so seldom. I declare I'd just like to see that little feller's eyes when he sees the tree I'd make for him. I'll make it just as near like the one I made four years ago as two peas. I'll—say, will you help?"

"Yes, I will, by George!" cried Scott. "I'll hold this message over, if I lose my job for it! Shall we tell Jim?"

"Of course, let's tell Jim. Half the fun will be in letting old Jim help. Come on over to the store. I'll open her up and get the things out now."

"No, I'll go and cut the tree."

"Do you think you'll get the right kind? Mine was about so high and straight. You'll find some over by the—"

"I know—I know," said Scott, hastily. "No wonder Carson ain't a success," he thought, as the other hurried away. "Nobody ever is that's so took up with other people's troubles and other people's Christmases the way he is. You've got to stick to business and look out for Number One, and let the other fellow go to the devil, if you expect to make anything in this world. Lord, what a God-forsaken spot a mining-camp is on Christmas!"

He took an axe and a lantern. Then he went out and made his way to the graveyard to cut his Christmas tree.

II

Carson went out to get Jim. He found him in his little cabin, coughing and shivering in spite of the mild weather. His gentle eyes were feverishly bright, and his thin form was almost racked to pieces with every cough.

"Come in, Carson, come in," he said, eagerly. "It does me good to have you drop in. I was thinking I shouldn't get much sleep to-night. I'm too excited over Mollie's coming, and the baby's. I call him the baby yet, but he must be too big for that now."

STRIKE AT THE "BILLY BOWLEGS"

"You sure she'll be here to-morrow?" asked Carson.

"Sure! I've heard from her. She came on the train I expected, and hurried right on. Fresno has seen her. He got to camp to-night. He said she was well, and that she was pretty. Fresno actually said that, and he said the boy was a peach. He said she was so excited at getting here that he never told her about the strike. He said he guessed she'd know soon enough."

Carson's mind was filled with her. Young, pretty, and excited with the idea of seeing her husband — that poor, shrunken figure, hovering over the fire with his transparent fingers held out to the blaze.

"Seems to me you're worse to-night, Jim."

"Me? Of course I'm worse! Why shouldn't I be worse to give poor Mollie the most forlorn welcome possible. The doctor says this mining is killing me. I got better the first few months I was West. Then I took to mining, and I've got worse ever since. The climate alone would cure me if I could take advantage of it."

"I believe it would. It's cured worse cases than yours. Say, Jim, I've been thinkin' about goin' back East to my folks, only I don't know of anybody I'd care to leave the store with. Do you believe you'd like that job?"

Jim got up and crossed over to Carson with a fierce, wild glare of eagerness in his face.

"*Like* it?" he cried. "*Like* it? Why, old man, I'd rather have it than anything else in the world! Do you mean it? Oh, to think that I haven't brought Mollie and the baby here to starve!"

He wrung Carson's hand. Then he sat down and buried his face in his hands. His thin shoulders heaved.

"Look here, Jim! Don't do that. I've got another little scheme. Scott's gone out to cut a Christmas tree, and we thought if you'd brace up and help us we'd rig up a tree for the boy to-morrow that would please your wife. We must show her that she hasn't come to the jumping-off place."

Young. Pretty. With blue eyes, perhaps, like Katie's. And the boy? He wondered if he would remind him of Billy.

"Jim's exclamation of delight roused him.

"Come on, then! Let's go over to the store and get the things. I—I've got my mind about made up as to how we can trim the tree. Can you string cranberries?"

"Can I? I've strung barrels of them back in the State of Maine. And pop-corn, too."

"That's the ticket!" cried Carson, delightedly, hunching his great form along beside Jim's slight one. "I declare, you look better already, old man. I believe you'll get well in spite of all we can do to kill you off."

Jim laughed gleefully. The sound fell curiously on the sullen air of Cactus Camp. A door opened as they passed, and a woman thrust her head out to see who could feel like laughing at such a time as that. Her grim face relaxed when she saw Jim. She thought, because of his wife and child, that he was the only one in Cactus Camp so privileged. "And God knows he has little enough call to laugh," she said to herself, as she closed the door.

"Who was that?" asked Carson.

"It was Peters's wife. Poor woman! She knows Peters is off somewhere with the boys drinking, and trying to bolster himself up for the fight."

"What fight?" asked Carson, apprehensively.

"I mean the strike," explained Jim.

"Oh," said Carson.

They came to the store.

Carson unlocked it and entered. He struck a light and held it over his head. His rugged face had an expression of satisfaction upon it which made it good to look upon.

He poked around, scooping up cranberries and sugar; measuring out cochineal; doing up eggs and pop-corn; counting out candles, and even taking out of the showcase a box of store-candy from Denver.

Then, burdened with all these breakable articles, the two men made their slow way over the uneven ground back to Jim's cabin.

STRIKE AT THE "BILLY BOWLEGS"

As they passed Peters's cabin they heard the sound of sobs.

They crept up to the door and listened. The thought was in the mind of each that possibly Peters had come home drunk, and was beating his wife. The thing had happened before. But this time they only heard a woman's sobs and broken prayers to the Virgin. Her wretchedness over her situation seemed complete enough without the thought of Peters's return, who, as ringleader in the strike, was at the bottom of all this misery. There were some who declared that they could hold out six months.

Carson and Jim crept away from the door, and reached Jim's cabin in silence.

They found Scott waiting for them with a Christmas tree which exactly met Carson's requirements.

"That's it! That's the ticket!" he exclaimed. "I couldn't have got a better one if I'd gone after it myself. Say, Jim, isn't that a beauty? Now nail these here things together, Scott, to make it stand up, and just trim it off a little on the bottom while

Jim strings the cranberries and I pop the corn. This is goin' to be exactly like—like one I made a long time ago."

Jim was too much overcome at the sight of those two big men working about his little cabin, making a Christmas tree for his boy, to begin the task that Carson had set for him.

He sat helplessly watching them. The blows of Scott's hammer deafened him, and the odor of Carson's pop-corn filled the room.

"I declare, boys—" he began, brokenly. Tears gathered in his eyes.

"Say, Scott," broke in Carson, nervously, " guess who is at home with her children, cryin' and sayin' prayers to the Virgin instead of eggin' the other women on the way she did in the last strike?"

"Not Em Peters?"

"The same. What's got into her, do you s'pose?"

"I guess she got enough of strikes two years ago," said Jim, "when she and her children nearly starved to death before the thing was settled. She was the worst of the lot that time. She was out-doors day and night, egging the men on and urging the other women to back their husbands up. Peters was a quiet enough sort of man then. He had to be egged on all the time. Now, I guess, she sees the result of her work. It's Peters that started this strike. But he is keeping up his nerve on whiskey. I don't believe his sand will hold out."

"Is that so?" said Scott, pricking up his ears. "I thought he was a good deal of a devil."

"Not a bit of it. If anybody could get on the right side of Peters, I believe he'd back down, and the rest would follow like a flock of sheep."

"You don't say so."

"It's too bad there isn't somebody in camp that's got the gift of gab, who could talk to them. I wish the boss knew as much of the men as I do. He'd send out somebody who could handle them like putty. But they think he will show fight, and they are fixing to play a waiting game. They don't believe he will call out the troops the way they did at Leadville, because he is too anxious to mine. They expect to stand a siege."

Carson and Scott looked at each other.

"Well, what if he does order out the troops?" said Carson.

"Oh, he won't," rejoined Jim. "At least not till after he's talked and talked. They've struck before. They know what to expect. He'll *have* to give in sooner or later. They are not afraid of being killed. That's always the last thing, you know. They will have plenty of time to brace up for that."

"Well, what would happen," said Carson, in a curious tone, watching the gentle expression on Jim's girlish face, "if the boss got troops ordered out the first thing and didn't give them any time to think or brace each other up or to talk? Suppose he just fired on them. What then?"

"Oh, they'd be killed — lots of them," said Jim, quietly — "lots of them, that were just led into it by others. Lots of the comparatively innocent. These men can't fight. They can only throw stones and shout. They are not brave. They are stubborn.

Oh, yes, they would be killed, but — there wouldn't be any more strikes."

Jim's gentle voice fell with an ominous note upon the ears of Scott and Carson, with the knowledge of the suppressed telegram in their minds.

"Too bad to have your wife come just now," said Scott, to change the subject, "when you are out of a job."

"Oh, I am all right," said Jim, placidly, while a beaming smile broke over his face. "Carson, here, is going back East to see his folks, and he is going to give me his job. Did you ever hear of such a piece of good-luck for Mollie and the baby?"

Carson waved the corn-popper at Scott over Jim's unconscious head. Scott looked at the happy light on Jim's face as he busied himself with his homely task, and then at Carson's huge shoulders and shaggy head. He remembered that glimpse of his heart Carson had given him but a few hours before. He thought of the arrival of Jim's wife and boy. He thought of Emma Peters weeping alone in her cabin, and of the telegram hanging over the heads of them all.

If only somebody who could talk—somebody who could do it right—could tell those men what was going on in their midst. A sudden thought thrilled him. He wondered if he could do it? He was not gifted with speech, but was speech necessary?

"Say, you fellows," he said, getting up and stretching himself, "you will have to finish this blooming thing by yourselves. When it's done, I'll come and look at it and pat you on the heads if it is well done. And, Jim, I'll bring a present to hang on it for Mollie and the baby."

III

As if afraid that his courage might waver, Scott hurried to Pedro's, where he knew he should find Peters and most of the others. Some of the miners, the best or the most timid, were at home with their wives. The most of the men at Pedro's had no wives. Peters was an exception.

He was greeted with a shout as he entered.

"Have you come to join us?" two or three asked.

"Yes," he said.

"Good! Good!" they said, crowding around him.

"In a drink," he added, with a grin.

He caught them just right. They fell back with a roar at his joke and demanded to hear his order.

"Lord knows I need it," he said, tossing it down and ordering another. "I've just about had the ghostly jim-jams."

He fixed his eye on Peters as if addressing him, and Peters, like all weak natures, yielded to his influence and moved nearer.

"What's the matter?"

"Why, Carson discovered that it was Christmas eve a little while ago, and what do you think he did?"

"Don't know. Tell us."

"By the way, did *you* know it was Christmas eve?" he demanded suddenly of the flushed faces thrust closely around his.

They shrank back and looked down, nervously picking at their coat-sleeves.

STRIKE AT THE "BILLY BOWLEGS"

Christmas eve! The words brought back memories to all of them.

"You know, Carson is a big, soft-hearted old codger, that would walk two miles to keep from hurting even a buzzard, and I guess he got to thinking what an infernal Christmas the women and the kids in this God-forsaken camp are going to have to-morrow. No Christmas trees, no toys, no presents of any kind— Why, blame it all, it kind o' gets *me* to think that *I* ain't going to have a Christmas present. Not one person in all this world is going to think of me to-morrow. I tell you, boys, those of you who've got wives here, and who've got sweethearts anywhere, better be sorry for poor devils like us who haven't got even the love of a prairie-dog.

"Well, as I was saying, Carson came over and sort of opened up on me about his Christmas four years ago, when he had a wife and baby. He was rigging up the bloomingest kind of Christmas tree for the kid, all hung with strings of cranberries and pop-corn, and lighted with candles, fit to make a child's eyes pop out of his head to see it, and right in the midst of it the little feller took sick and died."

Scott paused and searched the faces of the men. They were listening intently. Not one was drinking.

"Didn't know Carson was married," said Peters.

"Neither did I. Or, at least, I'd forgotten. But when he mentioned it, I remembered that the mine was named after his boy."

"The Billy Bowlegs!" exclaimed Fresno. "It was, sure enough. Carson and his brother discovered it. They worked it for years, and the week after they sold out it began to pan out like blazes. If Carson had held on a week longer, all this stuff would 'a' been his, and *he'd* be the boss we are fighting."

"You'd never fight Carson when you hear the rest," said Scott, shaking his head.

"Go on," said Peters.

"Well, the upshot of it was that he was down on his luck with this here coming of Jim's wife and baby, and the tough luck of her coming on Christmas Day and finding

STRIKE AT THE "BILLY BOWLEGS"

the strike on, and the camp in such a fix. Carson takes other people's troubles to heart as if they was his own. He declared Jim was goin' to die, and wasn't it the meanest sort of lines for that poor little woman to have saved up for so long, and for Jim to have saved up, and skimped, and not eaten half he ought to, just to get her out here quicker, and then to have her come to this. I just wish you could 'a' heard Carson take on. You'd 'a' thought this strike was the worst thing that ever struck the world since the flood. You'd 'a' thought Mollie and the baby were his own wife and boy he was grievin' after. You'd 'a' thought we were all a set of cannibals or heathens to do anything towards carryin' this fight on while that poor little woman was comin' all the way from the State of Maine, headed straight for the perdition of Cactus Camp. I declare I felt as though there was blood on my hands. Poor Carson! Anybody 'd think Jim's wife was the only woman to be thought of. I guess you've got wives, some of you fellows, that haven't got a much pleasanter Christmas to look forward to."

He stopped and glanced around. The men looked down in confusion and kicked at the sanded floor.

"The little woman is pretty," said Fresno, suddenly. "I saw her yesterday. She was 'most wild to get to Jim. And the boy—say, but he's a hummer."

"Did you tell her about the strike?" asked Peters, anxiously.

"No, I didn't. I lied handsome. I said Jim was well and the camp never better. Poor little thing! She'll find out soon enough. Mesquite Dan is bringing them in his wagon. They'll be here to-morrow."

A dead silence fell. None of the bravado of an hour ago was visible.

Scott saw his advantage and hurried on.

"Well, sir, after a while, Carson got so worked up that nothing would do but he must rig up a Christmas tree exactly like the one he was making four years ago to-night for his own little feller, and have it ready to-morrow for Jim's. He sent me after dark to the graveyard to cut one of those baby pines, and he opened up the store and got the candles and stuff, and he's over there at Jim's now, trimmin' it!"

"Well, I swan!" said Fresno.

"Let's go and see him," said Scott. "We can look through the window."

"All right," they exclaimed.

There were fourteen of them.

"Jim won't live six months," said Peters, sullenly.

"Oh, I forgot to tell you that Carson is going back East 'to see his folks,' he *says*, but he has given Jim his job."

A chorus of exclamations greeted this announcement. Those men were quicker to recognize and appreciate a fine deed than their more civilized brothers. Smiles played over their countenances. Gentler expressions came into their eyes. Carson was a hero among them. If he had been there they would have given him three cheers.

"Jim 'll get well now that he's out of the mine," said Fresno.

Scott purposely took them by Peters's cabin.

"Hold on. What's that?" asked Fresno as they came near.

STRIKE AT THE "BILLY BOWLEGS"

"Oh, it's only one of the women takin' on over the strike," said Scott, carelessly.

When they reached Jim's cabin, they stealthily crept up and peered in at the window one at a time, then fell back into the shadow. Scott cautioned them to be quiet or they would get caught.

Peters was the last. As he fell back, he drew the back of his hand across his eyes, and blew his nose tempestuously.

"Stop that, you old fool," said Fresno, savagely, treading on his foot.

"I've got a cold," explained Peters, in a hoarse whisper.

"Well, take it out and lose it," commanded Fresno. "We don't want to get caught in this. But say, boys, wasn't it pretty? It looked just like the one my mother made for me once when I was a little chap."

"I remember I had one, too," said Peters. "My kids never saw such a tree."

"Why don't you make 'em one?" asked Scott.

"With this strike on?" demanded Peters. And as they again neared his cabin he added, "Does that sound much like Christmas to hear Em goin' on like that?"

"Oh, let's give up the blamed thing," said Fresno. "What's the use of letting Jim's wife be the only happy woman in camp to-morrow. Go on in, Peters, and tell Em to dry up, and let's all have a Christmas!"

Scott held his breath.

Peters hesitated, then turned in at his doorway.

"Hooray!" cried Scott. "Come on, boys. I'll open up the office and telegraph the boss that the deal is off. You'll all get more out of it in the end."

Carson and Jim heard the noise and came out, running. They heard the news, and they, too, joined the shouting. And it was not long until the whole camp was out in various stages of undress to join the others in shouting as loudly because the strike was off as yesterday they had shouted because it was on. They were indeed like sheep.

Scott lighted up the telegraph-office and wired the message to Boston before they had time to reconsider.

STRIKE AT THE "BILLY BOWLEGS"

Carson was called on to unlock the store for the second time that night for Christmas-tree supplies. The women stood around with shawls over their heads, unable to comprehend the sudden and wonderful change. Christmas! It meant nothing to them.

"Just wait," the men said.

Mysterious noises and lights kept the camp uneasy all the rest of the night. Hard faces relaxed and grew tender over the sweet old-fashioned thoughts which came thronging in with memories of long-forgotten childhoods.

When Christmas morning dawned in Cactus Camp, and Mollie and the baby came, they were met by Jim and Carson and Scott, and taken post-haste to see the tree.

Carson slapped his leg every time he thought of it. The boy wouldn't even leave the tree long enough to eat his dinner. Carson had the satisfaction of seeing just "how a little feller's eyes would shine at the sight of that tree."

By night there was not a tree left in the graveyard.

A WOMAN OF NO NERVES

A WOMAN OF NO NERVES

Aunt Ann lived in Deerfield, Massachusetts, and after I had seen her I knew that she could have come from nowhere else. There was something about her which suggested Deerfield. Her appearance indicated long, slow streets, noble, dignified trees, and green meadows full of a benevolent content. She could not have come from Poughkeepsie or Hackensack or Yonkers. These names were not in keeping with Aunt Ann, who was mild-eyed, capacious, contemplative.

Aunt Ann was unmarried, and must have been about thirty-eight. She still possessed the delicate bloom on her plump cheeks with which New England loves to bless her daughters. Of ample but not unsightly proportions was Aunt Ann, with the gentle, direct gaze of an Alderney cow — a cow in a clover pasture, imbued with the meditative atmosphere of Deerfield, Massachusetts.

She was almost beautiful, and there was a Sabbath calm in her presence which led one's thoughts, perhaps, not quite to religion, but at least as far as ethics.

The first time I saw her she entered the room where I had been ill, and simply looked at me out of the dimness—a steady, steadying look which got me my bearings and gave me something tangible to hold on to. She had a bovine effect upon me which was most agreeable. Then she went away without whispering to mamma to tease my curiosity, and when she spoke she spoke aloud. I heard her laugh in the next room, a deep-throated, comfortable laugh, with a note of contagion in it, causing a shadow of a smile to pass over my own face.

For days I did not see her again, but I heard her step in the hall, slow, measured. I knew that her shoes were comfortable from their almost inaudible creak, as if her whole weight rested easily upon each sole as she walked.

I had so dreaded the advent of this maid-

A WOMAN OF NO NERVES

en aunt from New England that I was agreeably surprised thus to learn to like her through closed doors. I found myself longing for the calm personality which I supposed must accompany these symptoms of pleasantness. When she came, her gentle touch quieted me. Her even, steady tones were inexpressibly soothing. I liked to feel her smooth the pillows.

But as I grew slowly better I discovered that there was a board in the floor which suddenly developed an exasperating squeak, a reluctant, long-drawn sound which nearly drove me wild. Aunt Ann never failed to step on that particular board. I found myself, from the moment of her entrance, dreading to have her reach, tread upon, and release that complaining bit of the floor. In vain I reminded her in advance. She forgot, and never seemed to hear it. She could rock on it for an hour with no hint from her own sleeping nerves that she was driving the more sensitive frantic. And then, to my despair, for I honestly loved her, I discovered that Aunt Ann never could sit very long without jingling two of her rings together or fingering her bunch of keys or tapping her thimble on wood. When she was a child I suppose she wrote with a slate-pencil which — but why refer to a sound more horrible in my ears than the wail of a lost soul?

She came in one morning to sit with me, with a roll of white cotton cloth in her hand, which she quietly unfolded and began to tear into long, narrow strips. With an angry, snarling rasp the threads yielded, and Aunt Ann, with her plump arms separated to their fullest extent, complacently tore and tore and tore that awful cloth until, in my weak state, I burst into a flood of childish tears.

"What is the matter, dear?" asked Aunt Ann, dropping her work and coming towards me. She stopped and looked back.

"Was I rocking on that board?"

I shook my head and began to laugh hysterically. It was so absurd of her not to hear those dreadful sounds.

"It was—it was the cloth," I managed to say. "It makes me shiver so. How can you stand it yourself?"

"Such trifles never disturb me," she answered, helplessly, "and I can't tell when you are nervous."

"But I am always nervous. It is not because I have been ill."

"It is such people as you who are always having nervous prostration," she answered. "You burn up with inward fires. It is very bad for you. I never was ill in my life."

She was on her knees beside me, smoothing my hair and holding both my thin hands in one warm, steady palm.

"You are very good to me," I murmured, gratefully.

"You remind me of some one," she said, with a far-away note in her voice as of one who recalls something out of a tender past. For reply I only pressed her hand feebly. A strange, fluttering look swept the placidity from her face. "You are so like him!" she whispered, suddenly, with the subdued excitement of those who never yield their self-control except when it is wrested from them by a force from within.

We had been strangely drawn together from the first, Aunt Ann and I, for, notwithstanding our wide difference in temperament, there was a keen bond of sympathy between us, so I was not greatly surprised that she should speak to me of the lover of her youth, whose existence we sometimes had doubted and whose name we never had known.

"When he was ill at our house," she proceeded, with a pathetic idea that I knew of whom she would speak, because he had been so great a factor in her life, "so long ago, we were talking one morning while I was working on my china-painting, when something—I never knew what, grown man that he was—made him burst out crying like a child. He was so weak, dear, and had been so very ill."

A note of apology for her lover's lack of strength crept into Aunt Ann's voice.

"Of course; I know. When one has been ill—" I said.

She stroked my hand gratefully.

"I was young then, and it frightened — choked me. I ran to him and dropped the plate I had been burnishing. I have it yet, broken in two pieces. I knelt down by the couch and he took my hand, just as you did, and he pressed it with his poor, weak fingers, just as you did. Oh! you are so like him! But the pressure meant only gratitude for my kindness. He had been ill and helpless so long, and I had tried to amuse him and occupy his mind. He was all fire and nerves and cleverness. I had been of much use to him. I seemed so strong and well beside him. His weakness went to my heart. Women's hearts are weak, I think, dear. But he was only grateful, while I — it was more with me," she said, quietly. "He used to say that I soothed him, that it rested him to look at me. He said I would be just like my mother at her age. He used to call her his benediction.

"But I had given him so much out of my simple life. I did not know that he was giving me only gratitude. Such an awful sensation came over me when I realized it! He had been talking of a woman to understand him, the kind he needed, and I thought he meant me. But he was thinking of some one else. All at once I felt it, felt that he could love, did love — somebody, and — I never have felt anything since. I put all thought of him away from me then, and on the surface everything went on as usual. But there was a difference. Sometimes I think he saw the change in me and wondered, for I remember how, whenever I looked at him, I always met his eyes. Then he went away. I never saw him afterwards. We heard from him, of course, for he was so grateful. He had come with letters to us from a dear friend, and had been taken ill at our house, and, as was only natural, mother cared for him through it all. He loved my mother so beautifully.

"His letters were very sad sometimes, so I used to think that his love for this other woman was not a happy one. He mentioned me, too, kindly, almost tenderly, but I never allowed myself to think of—what might have been. That is all. It is little enough, isn't it? So very little to last one's heart a lifetime?"

"Dear Aunt Ann," I said, remorsefully, "I'm so sorry I was silly enough to cry, and remind you, and bring it all back. *Please* to rock on that squeaky board and tear some more cloth. I—I *like* it!"

Aunt Ann laughed comfortably. Hers must be a wholesome, healthful nature to be able to laugh like that.

"It has done me good to speak of it. I have been tempted to tell you ever since I came. You are so like him."

She left me and sat down again, carefully avoiding that board with a deprecating glance at me that made us both smile.

And she never knew that after she had rolled up the objectionable cloth she sat tapping the arm of the chair with her thimble for twenty-three minutes.

When I became well enough to come down to the library, and Aunt Ann was still with us, a friend of mine, the night editor of *The Sun-Dispatch*, dropped in to see me. The day was raw and cheerless, and he begged me for a cup of tea.

"You poor little thing," he said, as I busied myself with the cups. "You look as if the draught through a keyhole would blow you away."

He was a bundle of nerves himself, with a thin, eager face, iron-gray hair brushed back from sunken temples, and a smouldering fire in his eyes which told how his strength went. When he sat down he had an alarming way of sinking into his clothes as if some day he might disappear from sight altogether. I always was agreeably surprised when he emerged.

He stretched a shaking hand out to take his tea, and laughed apologetically.

"It might be the result of cigarettes or intoxicating liquors or opium," he said, in gay self-derision. "What a pity it is so respectable a thing as lack of sleep."

I reached for the tongs to stir the fire, and, setting them down carelessly, they fell with a frightful clatter on the marble hearth.

He gave an expression of exasperated impatience.

"How cross you are!" I cried. Then perhaps the shock to my own nerves, combined with the look of physical pain between his drawn brows, gave me an excuse to be absurd, and tears started to my eyes.

"If you cry," he said, setting down his cup with much deliberation, "I shall go home and shoot myself, and I shall keep on shooting myself until I am stone dead.

A WOMAN OF NO NERVES

There! That's right—laugh. I deserve the shock to my vanity which it gives me to see you laugh in pure delight at the idea of my demise."

"You do, indeed, deserve it. I never knew before that you were ill-tempered."

"Heaven forefend!" he exclaimed, bringing his head out of collar to be more impressive. "Never, with your fine sense of discrimination, make the mistake of confounding ill-temper with raw nerves. One is an infirmity; the other an ailment."

"The result is the same, and both can be cured," I answered, emphatically.

He dropped back into his chair with a sigh.

"I wish I believed that, and, if true, I wish I had known it twenty years ago."

"Why, would you have established a sanatorium?"

He smiled.

"I would. A permanent one, for my own exclusive cure, and with only one woman needed for the entire staff of physicians."

"What do you mean?"

"I gave up the only woman I ever loved

because she was born without a single nerve, and because she never knew when she was driving me to distraction. I had the courage then to do it. Life was all before me, and I did not know then that love comes but once. I did not know how little real love there is in the world. Oh, if men only knew, they would be kinder to it when love does come! Nor did I know how much I myself cared. Without vulgarly formulating the thought, I had the vague hope that I would find another as sweet and pure and true, who could comprehend that side of me which she did not. But I never have found her. And now when I look back I do not call my folly courage. It was stupid ignorance and selfishness, and I have been justly and even generously punished."

I was interested at once.

"How did she fret you? What made her, if she cared for you?" I questioned, with hypocritical obtuseness, for my heart was wildly beating with the hope that it might have been Aunt Ann. Absurd, impossible, ridiculous of me, but what if—what if—

"She did not care for me. Perhaps in

A WOMAN OF NO NERVES

time she might have done so, but she never knew. Fortunately, I was the only one who suffered. The dear little woman!"

"Tell about it."

"What a delightful creature you are! What a gift to be so interested in people!" he exclaimed, coming into view again to put down his empty cup.

"Well, it was years ago. I often wonder if she ever married. I wonder what kind of a fellow she married. I hope he is good to her. I should like to kill him if he isn't. Years ago — but that makes no difference. I know just how she looks if she is alive to-day. She looks as her mother did then, and that would satisfy any man in his senses. If men chose their wives oftener with regard to the mothers' looks and characters, there might be more marriages which retain their flavor. Ann's mother was the most restful picture your mind's eye can conceive, with a sweet, wholesome, clean nature—a freshness, a dewiness about her which seemed continually to murmur of 'Green fields! Green fields!'

"I went to these dear people for rest after an exhaustive political campaign, and I never received more gracious hospitality or met with kindness tempered with such simple dignity. I repaid their courtesy by pitching headlong to the floor in a faint, which was the beginning of a tedious illness. I begged, I implored to be removed, but the stranger within their gates was to them a brother to whom they established themselves keeper. I never shall forget their kindness—never. Even when I was quite well, and wished to relieve them of an invalid presence, they would not permit it, and they allowed me to convalesce at my leisure, lying about the house on convenient couches, and being coddled in the most delightful way.

"It sounds perfect, doesn't it? You would think that the presence, the occasional presence, of a lovely girl — for their dignity was of so perfect a type that Ann's was only a rare and dearly prized advent— would lend an intoxication to invalidism. She was like a June rose, and I know now that I loved her.

"Sometimes I would feel myself almost

carried away, and that I must tell her about it, but one thing prevented me. Do you know that if she were in your place this moment she could hear that ice-wagon driving down this paved street at a gallop and not even look up? She could hear coal put into the furnace all day long and not shriek. If I ever commit suicide you will find that both the ice-box and the furnace are full. In fact, you needn't even look. You will know!

"Ann never worked as other girls did. She used to bring her work into the room, and there would always be some irritating, maddening noise about it. Tell me, do girls nowadays squeak and scrape things without setting their own teeth on edge, like the thought of lemon-juice?"

"No. American girls of to-day are all abundantly supplied with nerve ganglia."

"I am delighted to hear it. It is what makes them so keenly sensitive, so adaptable, so irresistibly sympathetic."

"In behalf of American girls," I said, bowing, with my hand on my heart.

"Well, to proceed. I was a careless fellow then, without realizing it, and I must have given their orderly natures a deal of unnecessary disturbance. Many shiftings have come with the years, and I am so impossibly neat now in these my old bachelor days that I wish with grim amusement that Ann could see a change of which she would so heartily approve.

"I used to go through a pile of newspapers and throw them in a heap on the floor. Then, when Ann came, I would begin a discourse on some subject in which we were both interested and which required all my nervous strength to talk about, and as she listened she would smooth every crease from those beastly papers, and fold them, oh, so neatly! Now you will think me a brute when I tell you that not even the added flush which came into her delicate face from stooping, or the beauty of her white hands, caused me for one moment to forget the exasperating noise of those detestable papers. It almost distracted me. The squeak of a chair or the rattling of a chandelier or the jingling of ill-balanced china on the sideboard never disturbed either of those two women.

"One day I begged her to come and talk to me, and she came with her work. She had on a blue gown made of some thin stuff —is it muslin or lawn that you wear in the morning?—oh, well, dimity then; with the ruffles edged with lace. I see you smile at my awkward attempt to usurp the prerogative of the fashion editor, but a man who never notices ladies' dress may sometimes keenly remember the simple frock which his sweetheart once wore, and in his memory she always wears it and always looks the same.

"I saw that she had a plate in her hand, for she did china-painting, and in my eyes did it very daintily. You will call me sentimental if I tell you that on this one she had painted little wreaths of forget-me-nots tied with pink ribbons. Forget-me-nots and pink ribbons! Doesn't it conjure the whole scene and particularly the nature, the soul of the girl? Would any one but Ann have painted out the bare whiteness of the plate with so idyllic a theme? Dear heart! dear heart! I had almost made up my mind to speak. We were talking impersonally of the need of a certain man for a certain woman, and each secretly, as I thought, adapting the abstract phrases to our concrete senses, when suddenly she took up an emery pencil and began to burnish the gold rim of the plate. Did you ever hear the sound? Get me a grindstone and any little knife you want sharpened, and I will form an accompaniment to this conversation such as followed the only attempt I ever made to make love to a girl."

There was an undercurrent of bitterness to his half-bantering account which convinced me of his sincere feeling. Occasionally he lapsed from his effort at self-detachment, and swerved to either extreme. When a man is really indifferent he keeps his self-control without effort.

He leaned forward with his elbows on his knees, sinking again almost out of sight with his face between his hands.

"Ye gods!" he shivered, "I can hear it now, rasping and scraping the feeling from my heart, and leaving a scar which is there

A WOMAN OF NO NERVES

yet. If it had been any other woman in the world I might have thought she suspected what I was going to say, and mechanically seized upon the only thing at hand to conceal her agitation. But not Ann. She was not only without suspicion of my feeling, but without any response to it. For when out of sheer weakness and distraction and uncontrol I gave way and burst into a flood of nervous tears, like an hysterical girl, Ann made the only hurried movement I ever saw in her. She ran to me, dropping her precious plate and breaking it right through a little forget-me-not wreath. She knelt down beside me and took one of my hot hands between both her cool, pink palms. I looked at her in amazement, undeniably wishing that it might mean more than mere womanly compassion, sweet though it was. But her eyes met mine without swerving—a friendly, steadying gaze which brought me to my senses like a dash of cold water.

"Neither said anything I can remember. Broken apologetic nothings, perhaps. She went back to her chair. I watched her pick up the pieces of the plate, to see if I could detect any sign of her recognition of the sentiment contained in the broken wreath of forget-me-nots. I wanted to beg her to give it to me, but I had a horrible fear that she would think me a fool. Youth is more afraid of the imputation of sentimentality than middle age, you see.

"But no, I think Ann regretted her spoiled work, although she was gravely sweet about assuring me that she did not mind. Ah, well!

"I wrote to them until Ann's mother died, then I wrote to her once or twice, but she never answered my letters. I have often thought of going back to Deerfield and looking her up. But I am afraid that the old fire is not yet burned out, and that if I saw her happily married, with her little children about her looking like smaller pink-and-cream editions of their pretty mother, or unhappily married, I should want to kill somebody—most likely myself."

"You two would never have been happy together," I said, with a secret sweep of joy that nothing pertaining to myself ever had given me.

A WOMAN OF NO NERVES

"Traitor!" he cried, raising his head and showing me eyes softened with emotion. "You said nerves could be lived down. How dare you destroy an air-castle called 'Might-have-been' with which I may solace my sleepless hours? If she only had cared! Happy? I tell you we *would* have been happy. Dear Ann! She might have been satisfied with me. She was not exacting nor difficult to please."

"Absurd!" I said, with a feigned heat which I was sadly afraid his keenness would discover, because it was so badly done. "Imagine her here in my place, you and her married. Come, draw your chair nearer; don't you love to put your feet on the fender? Now look into the fire and pretend that the woman in this chair is Ann—not any other man's wife, but yours. Now. You are nervous and tired. Hear this chair squeak? Isn't this the way she would do it? You would stand it about five minutes, then you would leap to your feet with an irritated exclamation, and perhaps rush from the house. Do you hear this? Isn't this a *jolly* squeak?"

"How more than unkind — how cruel, even, a sweet woman can be!" he said, slowly turning upon me a look full of reproach. He had followed my thoughts childishly, and let himself drift into a dream of Ann as his wife with the delicious abandon with which a tired body sinks into an easy-chair. "I expected you to laugh, but I did not think you would mock."

"Because I do not think you are in earnest in your reform," I said. "Would you be gentle and patient with her, and say, 'Dear Ann, please don't,' instead of glaring at her as you did at me?"

His face twitched as he tried to smile.

"I must be going," he said, standing up and buttoning his coat across his thin form with pathetic dignity.

A sudden daring resolve combated my fear of intruding and ruining myself forever both in his eyes and Aunt Ann's.

"It is pouring rain. Will you do me a favor? Heap coals of fire on my head, for I have been brutally unkind. Stay to dinner and help me entertain a difficult, almost impossible, guest. I am so nervous at the

A WOMAN OF NO NERVES

prospect that I feel as if I should fly. Have you any other engagement? We will dine early, at half after six if you like, so that you may go whenever you feel that you must."

"If you really need me—" He hesitated.

"Sit down and wait for me here," I said, rushing out of the room in a distracted fashion for fear he might read my tell-tale face.

"Aunt Ann!" I cried, bursting open her door and discovering her in a blue dressing-gown, doing her hair before the mirror.

She turned deliberately. She never jumped.

"Aunt Ann! have you a light-blue gown of any kind—anything thin and soft—a—a—lawn?"

"I have a last summer's dimity," she said, slowly, letting her pretty hair fall in its natural waves.

She went to the closet and brought the dress, holding it out before her and shaking its folds.

"It needs pressing."

"Not a bit," I declared. "Put it on quickly. Never mind if you catch your death-cold. It's worth it. And do your hair a little looser—the way I did it for you. What pretty arms you have! There is a man down-stairs who will be here to dinner. He is a nervous wretch who almost sets me crazy to watch him, so if you do any squeaking or creaking, or dare to touch a newspaper, or jingle your rings, you will see me go right through the ceiling. I simply cannot stand it. You will be a dear, and remember? Let me pin that for you. Where does this blue ribbon go? That's lovely. What a sweet old-ivory color this lace has! Now you look exactly like that portrait of your mother. Have you a handkerchief? There. Come on. Oh, how pretty you look! I am sure Mr. Dudley will admire you. Goodness, Aunt Ann, let go my arm!"

"Who told you his name?" she said, steadying herself by the banisters.

"Told me whose name?" Then I broke down. "He is in the library—Mr. Gilbert Dudley. He doesn't know you are here, but he loves you—he has always loved you. He has been telling me about it—the other side of your story, and you are going to marry him and be very happy. Aunt Ann, I'll pinch you if you faint. You've lost all your color. Rub your cheeks with your hand. No, harder. Here, let me do it. Oh, did that hurt you? Well, never mind. You look all right. Aren't you coming? Surely you're not going to back out now, after all my trouble. Now go right in. It will be over in a minute."

I got behind her and almost pushed her in. I could feel her body tremble. Aunt Ann trembling!

I heard an exclamation—from him.

"Ann!" he said, in a dazed way. "Ann!"

Then a deathly silence which seemed to last five minutes. What was the matter? Were they dead—both of them?

I did a dastardly thing. I looked through the crack of the door. He was standing before her, holding both her hands and looking down into her eyes with such rapture that his plain face was beautiful, and his bent form was straight, with a dignity which it never had worn before.

Such a look! I turned away with tears in my eyes.

A WOMAN OF NO NERVES

Twenty years!

I put my head in at the dining-room door.

"Don't serve dinner until eight o'clock," I said.

THE END

BY MARY E. WILKINS.

SILENCE, and Other Stories. Illustrated. 16mo, Cloth, Ornamental, $1 50.

JEROME, A POOR MAN. Illustrated. 16mo, Cloth, Ornamental, $1 50.

MADELON. A Novel. 16mo, Cloth, Ornamental, $1 25.

PEMBROKE. A Novel. Illustrated. 16mo, Cloth, Ornamental, $1 50.

JANE FIELD. A Novel. Illustrated. 16mo, Cloth, Ornamental, $1 25.

GILES COREY, YEOMAN. A Play. Illustrated. 32mo, Cloth, Ornamental, 50 cents.

A NEW ENGLAND NUN, and Other Stories. 16mo, Cloth, Ornamental, $1 25.

A HUMBLE ROMANCE, and Other Stories. 16mo, Cloth, Ornamental, $1 25.

YOUNG LUCRETIA, and Other Stories. Illustrated. Post 8vo, Cloth, Ornamental, $1 25.

Always there is a freedom from commonplace, and a power to hold the interest to the close, which is owing, not to a trivial ingenuity, but to the spell which her personages cast over the reader's mind as soon as they come within his ken.—*Atlantic Monthly.*

A gallery of striking studies in the humblest quarters of American country life. No one has dealt with this kind of life better than Miss Wilkins. Nowhere are there to be found such faithful, delicately drawn, sympathetic, tenderly humorous pieces.—*N. Y. Tribune.*

NEW YORK AND LONDON
HARPER & BROTHERS, PUBLISHERS

BY MRS. W. K. CLIFFORD

MRS. KEITH'S CRIME. A Novel. New Edition. With a Frontispiece from a Drawing by the Hon. J. COLLIER. Post 8vo, Cloth, Ornamental, $1 00.

An exceedingly graphic and readable story.—*Rochester Herald.*

The book is a notable one, and has many passages of great brilliancy and much skilful character drawing.—*N. Y. Sun.*

The significance of its motive, and the vivid way in which it grapples with the ultimate problems of human existence, are enough to give it a strong hold on thoughtful minds. . . . The story is a vivid record of personal experience, and, given a nature like that of Mrs. Keith, the consummation is entirely logical. There are many passages of thrilling interest, and, in spite of the painfulness of the theme, the book has a strange fascination.—*Beacon.*

LOVE LETTERS OF A WORLDLY WOMAN. 16mo, Cloth, $1 25.

There is abundant cleverness in it. The situations are presented with skill and force, and the letters are written with great dramatic propriety and much humor.—*St. James Gazette*, London.

AUNT ANNE. A Novel. Post 8vo, Cloth, $1 25.

There are in fiction few characters more consistently and powerfully set forth; in its way this piece of work is perfection. The study is so remarkable that it is hard to believe that it is not from life.—*N. Y. Tribune.*

NEW YORK AND LONDON
HARPER & BROTHERS, PUBLISHERS

BY ANNIE TRUMBULL SLOSSON

DUMB FOXGLOVE, and Other Stories. With One Illustration. Post 8vo, Cloth, Ornamental.

THE HERESY OF MEHETABEL CLARK. Small 16mo, Cloth, Ornamental, 75 cents.

The only criticism that can be made is one of eulogism, first for the perception of the artistic finish, and next for the pathos, tenderness, and grace employed in the illuminating of one great momentous truth. This book of Annie Trumbull Slosson's ought to give comfort to many a vexed and erring soul. It is a poem of the inner life.—*N. Y. Times.*

A charming little volume, quite unique in its conception and execution, and its ethical significance is no less noteworthy than its art.—*Boston Beacon.*

SEVEN DREAMERS. A Collection of Seven Stories. Post 8vo, Cloth, Ornamental, $1 25.

They are of the best sort of "dialect" stories, full of humor and quaint conceits. Gathered in a volume, with a frontispiece which is a wonderful character sketch, they make one of the best contributions of the light literature of this season.—*Observer*, N. Y.

Stories told with much skill, tenderness, and kindliness, so much so that the reader is drawn powerfully towards the poor subjects of them, and soon learns to join the author in looking behind their peculiarities and recognizing special spiritual gifts in them.—*N. Y. Tribune.*

The sweetness, the spiciness, the aromatic taste of the forest has crept into these tales.—*Philadelphia Ledger.*

NEW YORK AND LONDON
HARPER & BROTHERS, PUBLISHERS

www.ingramcontent.com/pod-product-compliance
Lightning Source LLC
Chambersburg PA
CBHW021803230426
43669CB00008B/611